VIBRATION PROBLEMS IN GEOTECHNICAL ENGINEERING

Proceedings of a Symposium sponsored by
the Geotechnical Engineering Division
in conjunction with the
ASCE Convention in
Detroit, Michigan

October 22, 1985

Edited by
George Gazetas and Ernest T. Selig

ASCE
1852
®

Published by the
American Society of Civil Engineers
345 East 47th Street
New York, New York 10017-2398

The Society is not responsible for any statements
made or opinions expressed in its publications.

PREFACE

This special technical publication constitutes the Proceedings of a two-session Symposium held at the ASCE National Convention and Exposition that took place in Detroit on October 22, 1985. The symposium was sponsored by the Soil Dynamics Committee of the Geotechnical Engineering Division of ASCE. The main objective of the symposium was to address geotechnical issues related to vibration problems other than those associated with earthquakes.

The first session, entitled "Analysis and Measurement of Machine Foundation Vibrations," was organized by George Gazetas. It contains 8 papers covering analytical, experimental and design aspects of machine foundations. The first five papers are concerned with procedures to predict amplitudes of machine-excited vibrations of shallow, embedded and piled foundations. Rigorous and simplified methods are presented, while small-scale experiments show the capabilities and limitations of such methods of analysis. The sixth and seventh papers describe two different designs of compressor foundations and present the results of field tests aimed at evaluating their performance. The eighth paper outlines a numerical formulation for assessing the effectiveness of active and passive vibration isolation by means of trenches.

The second session, entitled "Detrimental Ground Movement from Man-Made Vibrations," was organized by Ernest T. Selig, with assistance from Braja M. Das, Ray Meyer, Toyoaki Nogami and Robert D. Stoll. In the development of this program, many examples of problems caused by ground vibrations were discovered. A number of these cases could not be included because they were in litigation or still being studied. However, seven papers are contained in these proceedings that provide a variety of useful examples of vibration effects. Two papers examine settlements caused by pile driving vibrations. One paper presents observed ground vibrations from vehicles on projects where vibrations were a problem to persons or equipment. One paper presents observed ground vibrations from various sources including dynamic compaction, vibraflotation, pile driving, machine vibrations and vehicle vibrations. Another paper deals with ground vibrations from dynamic (impact) compaction operations. A sixth paper shows observed building response to construction blast vibrations. The final paper describes a study of potential detrimental effects on slope stability of vibrations from blasting that was performed to improve slope stability.

Special technical publications like this one are intended to reinforce the programs presented at conventions. As such they contain papers that are selected to be timely and support the program purpose. The papers may be controversial and are sometimes less comprehensive than papers accepted for the Geotechnical Engineering Division Journal. It is the current practice of the Division that each paper submitted for a special technical publication be reviewed for its content and quality. For these proceedings, each paper received at least two independent peer reviews. In accordance with ASCE policy, all papers published in this volume are eligible for discussion in the Journal of the Geotechnical Engineering Division and all papers are eligible for ASCE awards.

The following Soil Dynamics committee members or cooperating persons from the general membership reviewed these papers:

P. K. Banerjee	J. E. Luco
S. K. Bhatia	R. G. Lukas
J. Bielak	G. D. Manolis
J. L. Chameau	J. R. Meyer
P. Christiano	T. Nogami
M. C. Constantinou	M. Novak
P. Dakoulas	M. J. O'Rourke
B. M. Das	M. S. Power
P. A. DeAlba	C. E. Sams
R. Dobry	G. Schmid
B. J. Douglas	E. T. Selig
V. P. Drnevich	C. Soydemir
G. Gazetas	R. D. Stoll
J. R. Hall	I. Tadjbakhsh
F. Hand	J. L. Tassoulas
T. Kagawa	E. E. Vicente
E. Kausel	R. D. Woods
R. L. Kuhlemeyer	M. K. Yegian

Thanks are extended to the authors of the papers, for theirs is the biggest and most important job in this symposium. The editors would also like to express their gratitude to the many reviewers whose prompt consideration made it possible to meet the tight deadlines. Finally, thanks are due to Shiela Menaker who arranged for the assembly and printing of this proceedings volume.

George Gazetas
Ernest T. Selig
Editors and Symposium Chairmen

CONTENTS

PART I
ANALYSIS AND MEASUREMENT OF MACHINE FOUNDATION VIBRATIONS

PART II
DETRIMENTAL GROUND MOVEMENT FROM MAN-MADE VIBRATIONS

v

EXPERIMENTS WITH SHALLOW AND DEEP FOUNDATIONS

Milos Novak*, M.ASCE

Abstract: Field experiments conducted on vibrating foundations are reviewed and the main findings derived from their comparison with theoretical predictions are summarized. The experiments involved surface foundations, embedded foundations, single piles and pile groups under harmonic excitation and a hammer foundation exposed to shock loading.

INTRODUCTION

Design and analysis of machine foundations is an activity closely related to industrial development and is one of the oldest civil engineering disciplines in which the theory of vibration has been systematically employed. Over the years, various theories applicable to individual machine foundation types were formulated and refined but their experimental verification lagged these developments.

Experimental investigations can involve full scale foundations, small scale field experiments or very small laboratory models. The advantages of small scale field experiments are that the conditions of the experiments can be fully controlled, the excitation forces are well defined and the propagation of elastic waves is not obstructed by artificial boundaries as is the case in laboratory experiments.

In this paper, observations derived from small scale field experiments conducted over the years by the author and his associates are described and summarized. The experiments concerned surface foundations, embedded foundations, single piles and pile groups. In most cases the excitation was harmonic generated by means of a mechanical oscillator. One full scale experiment with an operating hammer foundation exposed to impact loading is also discussed.

The focus is not on the description of experimental detail, but rather on the lessons one can learn from the comparison of the theory with experiments.

SETUP AND EVALUATION OF EXPERIMENTS

Experiments with surface foundations were conducted on rigid steel test bodies assembled of different numbers of steel sheets and bases. The base areas were square and ranged from 0.5 m^2 (5.38 ft^2) to 1.5 m^2 (16.15 ft^2); the masses ranged from 970 kg (2139 lb) to 3770 kg (8313 lb). The soil was a very deep deposit of loess loam. Detailed data on these experiments can be found in Ref. 14.

*Professor, Faculty of Engineering Science, The University of Western Ontario, London, Ontario, Canada N6A 5B9

1

Experiments with embedded foundations were carried out using concrete bodies whose sizes and masses were similar to those of the steel foundations. They are described in Refs. 14 and 19.

The test foundations were exposed to harmonic forces varying with frequency by the quadratic law, i.e.

$$P(t) = m_e e \; \omega^2 \cos\omega t \tag{1}$$

in which $m_e e$ = the excitation moment resulting from unbalanced mass m_e and its eccentricity e, ω = circular frequency of excitation and t = time. The excitation was generated using a mechanical oscillator. The Lazan oscillator was used in most cases. The oscillator was connected to the test body. The oscillator and the test body were also used in pile experiments as is shown in Fig. 1 in which the oscillator is set for horizontal excitation and together with the steel test body, attached to a reinforced concrete cap for pile testing.

Nonlinearity

Changing the excitation moment $m_e e$, sets of response curves were measured such as those shown in Fig. 2. A typical feature of such response curves is the reduction of resonance frequencies with the intensity of excitation. This reduction can be seen even from one curve if the backbone curves, indicated in Fig. 3 by dashed lines, are constructed. These backbone curves indicate the variation of the undamped natural frequencies Ω with amplitude. They can be established for each individual response curve by intersecting the curve by a pencil of straight lines passing through the origin (Fig. 4) and by calculating the points of the backbone curve using the simple relation (15)

$$\Omega = \sqrt{\omega_I \omega_{II}} \tag{2}$$

Figure 1.   Steel Test Foundation With Lazan Oscillator Set For Horizontal Excitation of Pile Cap

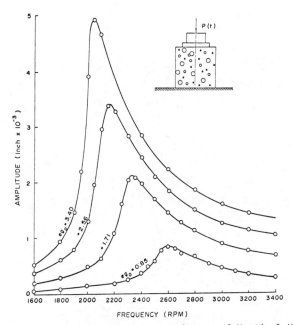

Figure 2. Typical Experimental Response Curves of Vertical Vibration
Measured on Surface Footing With Square Base at Various
Eccentric Moments ($eg_o = gm_e e$, g = gravity acceleration)

in which ω_I and ω_{II} are frequencies of the points of intersection. (Eq.
2 is strictly valid for single degree of freedom systems with frequency
independent stiffness and damping.) If the system is linear, the backbone curve is a straight line parallel to the amplitude axis indicating
that there is no variation in the natural frequencies with vibration
amplitude. In most cases, the backbone curve is curved which suggests
nonlinearity of the restoring forces. When there is a number of response
curves available, the individual response peaks should follow the backbone curve as shown in Fig. 4. Another indication of nonlinearity is
that the response amplitudes are not strictly proportional to the magnitude of the excitation forces.

The significance of nonlinearity is that it brings an element of
inaccuracy to the theoretical predictions and evaluation of experiments.
For rigid footings, the typical reduction in the resonant frequencies
with amplitudes amounts to some 20 or 25 percent which implies a considerable variation in stiffness. Strong variations in damping with amplitudes are also common. Thus, unless a nonlinear theory is consistently
employed, all experimental data should be reduced to one level of amplitude as a minimum measure.

The argument is sometimes made that in large foundations nonlinearity may be less pronounced than in small foundations because the

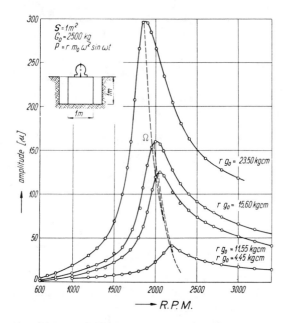

Figure 3. Set of Experimental Response Curves Measured on Rigid Foundation Exposed to Vertical Excitation; (rm_o=em_e, g_o=$m_o g$)

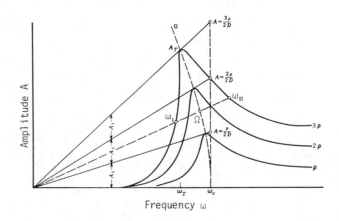

Figure 4. Construction of Backbone Curves of Response Curves

confinement conditions are more favourable. Experimental evidence of
this point would be useful.

SHALLOW FOUNDATIONS

Vertical Response

Surface foundations. - Presently, the elastic halfspace theory is most
often used to predict footing vibration. The applicability of this
theory can be verified, for example, by examining the variation of resonant amplitudes with footing mass and of resonant frequencies with
footing base area. Such an examination is presented in Figs. 5 and 6.

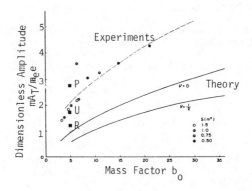

Figure 5. Variation of Vertical Nondimensional Resonant Amplitudes
With Mass Factor and Base Area For Surface Footings

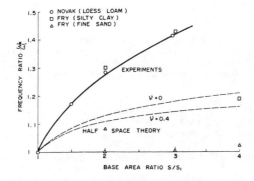

Figure 6. Relative Variations of Natural Frequencies With Base Area
With Constant Foundation Mass; Vertical Vibration, Surface
Footings

Figure 5 shows a comparison of experimental resonant amplitudes
with theoretical values plotted vs the mass factor

$$b_o = \frac{m}{\rho R^3} \tag{3}$$

in which m = mass of the footing, ρ = soil density and R = footing
equivalent radius. The theoretical values, shown in Fig. 5, were derived from the solution due to Awojobi and Grootenhuis (2), one of the
first "exact" solutions to the rigid body problem. All points shown were
measured at small amplitudes on loess loam, except for the point stand-
ing out of the sequence which was established on a deep deposit on sandy
clay. The trend of the experimental results follows quite well that of
the theoretical curves. However, the magnitude of the experimental resonant amplitudes is in all instances considerably greater (twice or
even more) than the theoretical values. Similar overprediction of the
vertical response was observed by Verbic (38).

Interestingly, the experimental amplitudes are quite close to those
calculated using the older approximate solution due to Sung (34) in which
uniform (U) or parabolic (P) stress distribution in the foundation base
is assumed in addition to rigid body stress distribution (R). For b_o=5,
the results obtained from the Sung solution are plotted in Fig. 5 for
the three assumed stress distributions and denoted by R, U and P, respectively. While Sung's solution, and similar older solutions such as
that due to Bycroft (4), are approximate because of the nonuniform displacement of the base, the mathematically more accurate rigid body solution may not be quite realistic either because the extreme stresses
along the base edges can not be maintained; they are likely to be dulled
and redistributed.

The discrepancy in the resonant amplitudes visible in Fig. 5 suggests that the vertical damping was about one half that predicted by the
rigid body-halfspace theory. Such a possibility is of great practical
importance because the halfspace theory generally predicts very high
values of damping for vertical vibration and even overcritical damping
for large foundations.

For layers of limited thickness, low values of damping are possible
when the frequency of excitation is lower than the natural frequency of
the deposit, but in the experiments presented in Figs. 5 and 6 the deposits were very deep and the resonant frequencies high.

The variation of resonant frequencies with base area established
for the vertical direction without changing the footing mass is shown in
Fig. 6 together with experimental results due to Fry (9). The frequency
variation is shown in a normalized form. The writer's results and Fry's
results obtained on silty clay with a reference area S_1 = 20.9 sq. ft
(1.94 m^2) agree very well but both indicate frequency variations far in
excess of those suggested by the elastic halfspace theory. The variations shown are more typical of a stratum than of a halfspace and suggest that some layering may be present even in an apparently homogeneous
deposit. The inevitable increase of soil shear modulus with depth due
to increasing confining pressure may also contribute to the discrepancies between the theory and experiments. For a layered medium, analyzed

as such, Tajimi (35) observed an excellent agreement between the theory and experiments in his study of a very large foundation of a shaking table. All these results suggest that a layered halfspace may be a better model than a homogeneous halfspace.

Fry's experiments obtained on fine sand and shown in Fig. 6 indicate no variation of the footing natural frequencies with the base area which does not appear to be consistent. This may be attributed to the difficulty of testing footings on the surface of a fine sand deposit. For such a deposit, the soil shear strength at the surface vanishes and the sand tends to flow out from under the footing along its edges during vibration.

Finally, the added mass effect, manifested by an apparent increase in footing mass, can be quite marked in the vertical vibration, particularly if frequency independent stiffness and damping constants are assumed. If this effect is neglected in the evaluation of the experiments, completely misleading results can be obtained for stiffness and damping constants. An evaluation procedure, suitable for both linear and nonlinear response curves, is described in (15). The added mass effect was also found by Verbic (38) even when he evaluated the theoretical response with frequency dependent stiffness and damping, i.e. with characteristics that are supposed to account for soil inertia. The apparent mass increase may be less severe for heavy full scale footings; in small scale experiments it is typically expressed by a coefficient ranging from 1.2 to 1.8.

Embedded foundations. - Embedment is known to increase both stiffness and damping of foundations which translates into increased natural frequencies and reduced resonant amplitudes. These effects are well documented, e.g. in Refs. 14 and 19. An example of experimental response curves affected by embedment is shown in Fig. 7. This figure indicates that there is a great difference between the effect of undisturbed soil and backfill, particularly with cohesive soils, with the efficiency of backfill increasing with density. In addition, the soil tends to separate from the footing under heavy operating conditions such as those encountered, e.g. in hammer foundations. Also, the embedment effect depends on confining pressure which reduces the validity of laboratory experiments conducted with very small bodies embedded in sand. Such experiments should be conducted in a centrifuge.

The factors mentioned complicate theoretical prediction of embedment effects. Yet, reasonable results can be obtained as will be shown in the next paragraph.

Response to Horizontal Excitation

Under horizontal excitation, a rigid footing with two vertical planes of symmetry undergoes coupled response characterized by horizontal translation and rotation (rocking) in the vertical plane. The response indicates two resonance regions with the second one usually suppressed due to higher modal damping and smaller efficiency of the horizontal force in the excitation of the second mode.

Surface foundations. - Examples of theoretical response curves

VIBRATION PROBLEMS

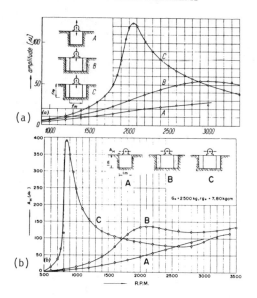

Figure 7. Effect of Embedment on Footing Response to (a) - Vertical
Excitation and (b) - Horizontal Excitation: (A) - Embedment
in Undisturbed Soil, (B) - Embedment in Compacted Backfill,
(C) - Air Gap Around Footing (14)

calculated with stiffness and damping constants due to Veletsos and
Verbic (36) are shown in Fig. 8. Significant effect of soil material
damping on the first resonance peak can be seen. This effect can be even
stronger with small experimental footings for which the radiation (geometric) damping due to rocking is particularly limited. Numerous field
experiments were conducted with bodies exposed to horizontal forces.
Invariably, the theory overestimated the resonant amplitudes by a factor
of two or even much more when soil material damping was neglected but
the prediction became quite good when material damping was included.
The typical value of $\tan\delta = 2\beta = 0.10$, where δ = loss angle and β = soil
material damping ratio, gave good results in many cases. An example of
the effect of soil material damping involving experimental resonance
amplitudes is shown in Fig. 9. (The data concerning this experiment are
given in Ref. 3.) For surface foundation ($\ell/r_0 = 0$) the theory overestimated the experimental values by several hundred percent when material
damping was neglected. Similar overestimation of resonant amplitudes
was reported by Petrovski (31) who observed a good agreement between the
theoretical and experimental frequencies but large discrepancies in amplitudes when material damping was not accounted for. The apparent mass
effect is not significant under horizontal excitation.

Embedded foundations. - Under horizontal excitation the embedment
effects on both the resonant frequencies and amplitudes are more pronounced than in the case of vertical excitation as a comparison of

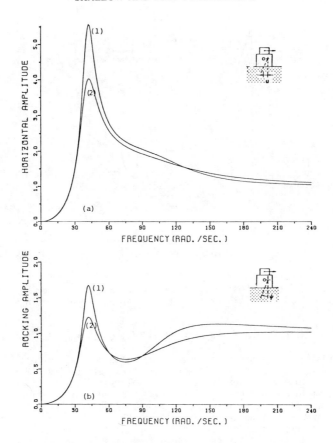

Figure 8. Typical Theoretical Response of Average Size Surface Footing
to Horizontal Excitation: (1) - Material Damping Neglected
and (2) - Material Damping tanδ = 0.1 included. (Base Area
160 sq. ft = 14.84 m²)

Fig. 7a and Fig. 7b reveals. Despite the difficulties of physical nature already mentioned, the theoretical predictions of the horizontal
response were quite good as can be seen from Fig. 9. For embedded foundations the contribution from soil material damping is less significant
because embedment generates considerable radiation damping (3).

Torsional Response

Torsional response (rotation about the vertical axis) is practically important because it is slightly damped and hence, large resonant
amplitudes can occur and the rotation can translate into severe

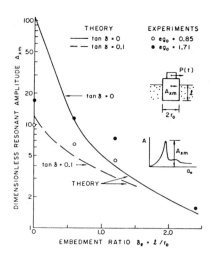

Figure 9.   Effect of Hysteretic Material Damping of Soil on Resonant
            Amplitudes of Surface Footings and Embedded Footings

horizontal translations along the edges of the footing.

       Significant progress has been made in the development of the theory
of torsional vibration cast in terms of linear elasticity. Veletsos and
Nair (37) elucidated the effect of soil material damping, Novak and
Sachs formulated an approximate theory for embedded foundations (28), a
problem analyzed with mathematical rigor by Apsel and Luco (1), while
anisotropy was considered by Constantinou and Gazetas (5). Still, the
theoretical prediction of torsional response using linear elasticity is,
on the whole, poor. This can be seen from Figs. 10 and 11 in which resonant rotational amplitudes and resonant frequencies obtained experimentally are compared with theoretical values. The experiments were
conducted using concrete blocks that were of circular, square or rectangular cross section respectively and embedded to different depths. The
theory used in this comparison is approximate but compares very well
with other solutions available (28).

       Figure 10 shows that in the experiments the natural (resonant) fre-
quencies are consistently about half those predicted theoretically and
this can be observed in surface footings as well as in the embedded
ones. As for resonant amplitudes, the theory overpredicts the observed
values by a factor of about two or more for the surface footings; for
embedded footings, the amplitudes are underpredicted in all cases by a
factor of two or three with the exception of the rectangular footings
embedded in undisturbed soil. It appears that all these discrepancies
can be attributed to slippage between the footing and the soil which is
almost inevitable in the case of torsion.

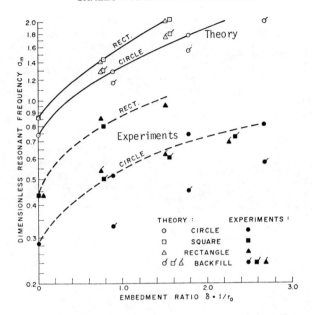

Figure 10.　Comparison of Theoretical and Experimental Resonant Frequencies of Torsional Vibration for Various Embedments (Theory in Full Lines, Novak and Sachs, 28)

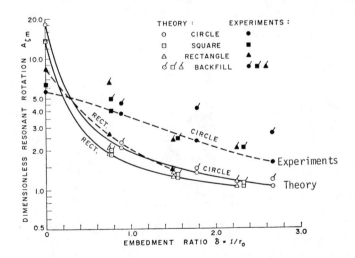

Figure 11.　Comparison of Theoretical and Experimental Resonant Amplitudes of Torsional Vibration for Various Embedments (Theory in Full Lines, Novak and Sachs, 28)

Slippage reduces stiffness and thus the natural frequencies for both surface and embedded footings (Fig. 10). At the same time, slippage generates frictional damping which far exceeds the small geometric damping of surface foundations. Hence, the resonant amplitudes of surface footings are reduced. For embedded foundations, slippage reduces their much larger geometric damping and consequently, increases resonant amplitudes. The rectangular foundation relies less on shear stress; hence, slippage is less severe in the undisturbed, stiffer soil and the agreement between the theory and experiment is better than in the other cases.

These observations suggest that a more realistic theory has to incorporate slippage. This was successfully attempted for surface footings by Weissmann (40). For embedded foundations an approximate approach allowing for a slippage zone around the footing and the slip in the base was formulated in (29). An example of the effectiveness of the latter solution is shown in Fig. 12. A reduction of the soil shear

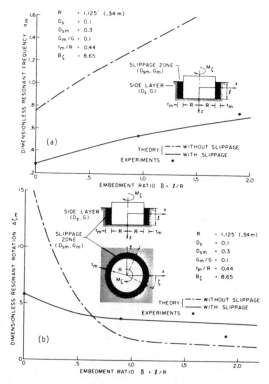

Figure 12. Comparison of Experimental Resonant Frequencies and Amplitudes of Torsional Response of Footings With Theoretical Values Calculated Either With or Without Regard to Slippage (Novak and Sheta, 29)

modulus in the slippage zone (G_m) produces a reduction in the resonant frequencies accompanied by a reduction of amplitudes for surface footings and their increase for embedded footings just as the experiments indicate. A more consistent theory for footings with slippage is needed, however.

PILE FOUNDATIONS

Many field experiments were conducted with piles and pile groups using an experimental setup similar to that used with shallow foundations except that the weight providing test body was mounted on the pile cap (Fig. 1).

Single Piles

In recent years a number of linear theories for dynamic analysis of piles have been formulated and formulae and charts were made available for pile stiffness, damping and deflections, eg. in Refs. 6, 13, 20, 23.

For machine foundations characterized by small amplitudes, linear theories seem adequate and indeed, experiments indicate that at small amplitudes, nonlinearity of the response of pile foundations is less pronounced than with shallow foundations (25). The reason for this is that pile properties, that are practically linear, contribute substantially to the total stiffness of the soil-pile system. Nevertheless, the theory has to be applied judiciously with regard to soil properties and some intuitive adjustments may be needed to account for pile separation. Otherwise, considerable discrepancies may occur between the theory and experiments.

Horizontal response. - The possible discrepancies are exemplified by Fig. 13 based on field experiments with small steel piles 6.1 cm in diameter embedded in a layer of fine silty sand, about 7 ft (2.1 m) thick, underlain by a more rigid layer of gravel and till (25). Figure 13 shows the horizontal response of a rigid test body supported by a single pile. In such a case, the pile stiffness relies heavily on the horizontal stiffness of the soil which diminishes towards the free surface as the confining pressure decreases and for sands vanishes at the surface. With a constant shear wave velocity, established in the field, the theoretical resonant frequencies, and sometimes even amplitudes, exceed the experimental values by 100% or even more. The discrepancy between the theoretical and experimental results was considerably reduced if a parabolic variation of shear modulus with depth was assumed for the sand layer and an allowance was made for pile separation from the soil. A separation along the length of two pile diameters gave good results. With these measures, the agreement between the theory, presented in (20), and the experiment became very good (Fig. 13).

It is worth noting that an equivalent constant shear modulus (or shear wave velocity, $\overline{V}_s$), i.e. equivalent homogeneous medium, can not produce satisfactory results. When it is chosen that the resonant frequencies (and thus stiffness) match, the amplitudes (and thus damping) get detuned and vice versa. Thus, the soil shear modulus variation with depth and specifically its reduction towards surface as well as pile separation must be accounted for if free ground surface without overburden

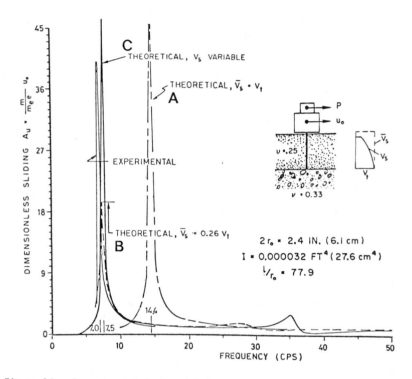

Figure 13.   Experimental Horizontal Response of Single Pile vs. Theoretical Predictions (20)

is present. Otherwise, both pile stiffness and damping can be grossly overestimated.

These problems are particularly topical for pin-headed piles tested in sand because such piles depend strongly on the topmost layer of the soil in which the shear modulus is quite compromised and poorly defined or absent and the confining pressure is negligible making pile separation under vibratory loading most likely. Figure 14 schematically illustrates those two factors. It is believed that large discrepancies between the theory and experiments reported by some researchers are to a very high degree attributable to those two factors. These notions are supported by full scale observations due to Gle and Woods (10) who found that the program PILAY (21), based on the theory presented in (20), overpredicted pile stiffness and damping by about 5 to 10 percent for soft to medium stiff clays, which is very good, while for granular soils, the overprediction was much larger. On the whole, Gle and Woods observed that the greater the ability of the soil to maintain contact with the pile, the closer the PILAY predicted stiffness and damping are to the experimental values.

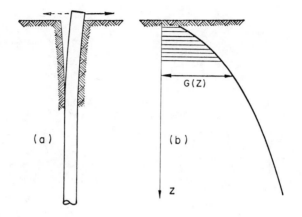

Figure 14. Schematic of (a) - Pile Separation and (b) - Reduction of
            Shear Modulus Towards Ground Surface

Thus, the problem is how to anticipate pile separation. Further
research is needed in this area. One approximate way of accounting for
the contact problems is to consider a weakened zone around the pile as
proposed in (29) and incorporated in the code PILAY2 (21). The main
effect of the weakened zone is to reduce damping at higher frequencies.

Vertical response. - Vertical stiffness and damping of piles depend
on the factors discussed above and also on the tip condition unless the
piles are very long. Floating (friction) piles feature smaller stiffness and larger damping than endbearing piles (16). This difference has
to be accounted for if a reasonable agreement between the theory and
experiments is to be obtained. This was learned, e.g., from the experiments (25) in which a very substantial overestimation of stiffness was
observed. Relaxation of the tip completely eliminated these differences (16). It is one point on which the old approach of the equivalent
cantilever completely fails because it yields greater stiffness for
floating piles than for endbearing piles.

Torsional response. - Torsional response (twisting) of single piles
is greatly affected by soil material damping. This is so because the
dimensionless frequency $a_0 = \omega R/V_s$, in which ω = circular frequency,
R = pile radius and V_s = soil shear wave velocity, is very low for piles
and consequently, very little radiation damping is generated in torsion
in analogy to shallow foundations. For small diameter piles the under-
estimation of total damping caused by the omission of soil material
damping can result in overestimation of resonant amplitudes by one order
of magnitude (Fig. 15).

Pile Groups

When the piles of a group are widely spaced their effect is just a

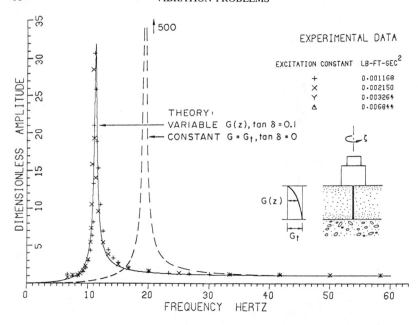

Figure 15.   Experimental Torsional Response of Single Pile Foundation
             vs Theoretical Predictions Showing the Effect of Soil
             Profile and Soil Material Damping (Novak and Howell, 1978)

sum of contributions from individual piles. Experiments with such groups
indicate the same sensitivity to soil profile and pile separation as observed with single piles. An example of this, based on experiments described in (25), is presented in Fig. 16. With a good estimate of pile
separation, two pile diameters in this case, and a proper soil profile,
a very good prediction was obtained using the code PILAY (21).

   Pile interaction. - When the piles of a group are closely spaced
they interact with each other. This interaction modifies both the group
stiffness, usually reducing it, and group damping, often increasing it.
Both are more strongly frequency dependent than the properties of single
piles. Significant progress has been made in recent years in the
development of the theories for dynamic pile group analysis and hence,
experimental verification of the peculiar features of pile group behavior and of the validity of the theories are of particular interest.

   An example of the examination of the vertical group action is given
in Fig. 17. The experiments were conducted in the field with a group of
four closely spaced steel piles 2.37 in (6.03 cm) in diameter and
11 ft (3.4 m) in length. The soil shear modulus G, varied with depth as
indicated in Fig. 17 and a weakened zone with shear modulus $G_m < G$ was
needed in the upper layers to account for pile separation. Soil material

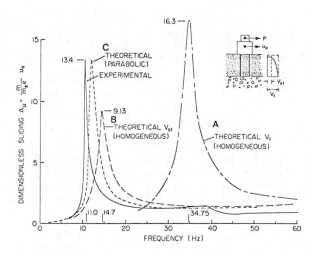

Figure 16. Comparison of Experimental Horizontal Response of Foundation Supported by Four Piles With Theoretical Predictions Calculated With (A) - Homogeneous Soil, (B) - Equivalent (i.e. Reduced) Homogeneous Soil Modulus and (C) - Parabolic Soil Profile and Pile Separation

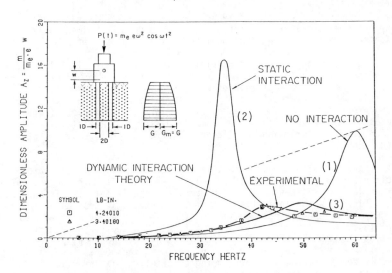

Figure 17. Experimental Vertical Response of Footing Supported By Group of Four Closely Spaced Piles vs Theoretical Predictions (Sheta and Novak, 33)

damping was $\tan\delta = 0.1$. The vertical response of a test body supported by this group was measured and compared with theoretical predictions made under the following assumptions: (1) no interaction was considered and pile properties were evaluated using the code PILAY2 (21), (2) static interaction was considered based on Poulos's static interaction coefficients (32) applied to both stiffness and damping and (3) dynamic interaction theory was employed as described in (33). It can be seen from Fig. 17 that the interaction effects cannot be ignored and that the dynamic interaction theory best predicts the reduction of group stiffness and increase in group damping. Another observation is that the weakened zone around the piles reduces the group effects but does not eliminate them.

For the vertical response of a large group of 102 test piles described below, similar observations were made with the additional need to account for the added mass effect with the very stiff piles used (8, 24) just as with shallow foundations. A few theories were tested but all required this adjustment.

Examples of *horizontal response*, both experimental and theoretical, of the large group of 102 test piles are shown in Figs. 18 and 19. This rectangularly arranged group was tested in the field and comprised small steel piles 1.05 in (26.70 mm) in diameter and 3.48 ft (1.06 m) in length. The soil was a special mix of sand and fly ash designed to have some viscosity and the same shear modulus as the surrounding soil. The test footings supported by the piles weighed 2,416 lb (1,096 kg) and 3,378 lb (1,532 kg) respectively. A detailed description of these experiments can be found in (8). The response to quadratic harmonic excitation acting horizontally was measured with different excitation moments given in Figs. 18 and 19 in Newton-meters (N-M). The individual response curves measured collapse quite close onto one dimensionless response curve indicating that nonlinearity is not very strong.

The theoretical predictions plotted in Fig. 18 are all calculated using the static interaction factors due to Poulos (32). First, it can be seen from Fig. 18 that when pile interaction is neglected (curve A), the stiffness is overestimated by a factor of about four and the damping is underestimated. Full values of static interaction factors (curve B) overestimate the reduction of group stiffness. Only tentatively adjusted static interaction coefficients (curves C and D) give acceptable predictions.

Figure 18 also suggests that if static interaction factors are applied because of the lack of dynamic interaction factors, they should be applied to stiffness but not to damping. The latter idea is supported by Fig. 20 which indicates that if the system stiffness changes in any way while the viscous damping constant is maintained, all resonant amplitudes are determined by a common tangent to the response curves passing through the origin. This theoretical relationship, proven in Ref. 15, holds for any linear or nonlinear stiffnesses as long as the mass is constant, the viscous damping constant, c, is invariant and the excitation is quadratic, i.e. proportional to ω^2. Such a common tangent is shown in Figs. 17 and 18 in dashed lines.

While Fig. 18 examines the ability of static interaction

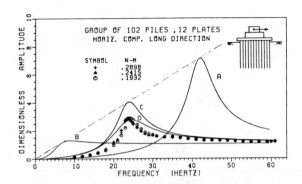

Figure 18. Experimental Horizontal Response Curve vs Theoretical Curves of Group of 102 Test Piles; (A) - No Interaction; (B) - With Static Interaction Factors Applied to Stiffness Only; (C) - With Interaction Factor of 2.85 Applied to Stiffness Only; and (D) - With 2.85 and -1.40 Interaction Factors for Stiffness and Damping, Respectively (24)

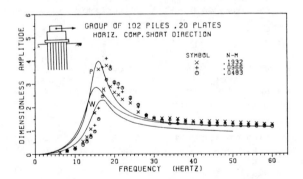

Figure 19. Comparison of Experimental Horizontal Response of Group of 102 Test Piles With Theoretical Predictions Based on: W - Waas and Hartmann Direct Dynamic Solution, K - Kaynia and Kausel Dynamic Interaction Factors and P - Equivalent Pier Concept (24)

coefficients to predict dynamic response of pile groups, in Fig. 19 various dynamic approaches are employed. The approaches included the direct dynamic solution due to Waas and Hartmann (39), dynamic interaction coefficients due to Kaynia and Kausel (12) and the concept of the equivalent pier (30) analyzed using the code PILAY2 (21). (This code was also used to calculate the properties of single piles for both Figs. 18 and 19.)

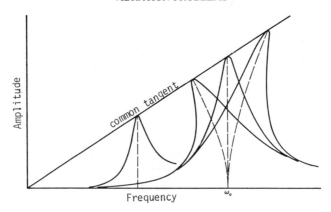

Figure 20. Relationship of Resonant Amplitudes of Viscously Damped
            Systems Differing Only in Stiffness (Quadratic Excitation)

       All the dynamic approaches yield quite reasonable predictions, far
superior to the static methods, which is very encouraging given the high
level of abstraction involved in the dynamic group analysis. Even the
much simpler model based on the equivalent pier, i.e. one equivalent
massive pile, gives a very good result if the geometric damping is reduced to 40 percent of the theoretical value.

HAMMER FOUNDATIONS

       All the above examples dealt with the behavior of foundations under
harmonic excitation. Hence, it is of interest to examine briefly the
ability of the theory to predict the response of footings to shock loading. Such loading is met most often in hammer foundations.

       The theoretical analysis rests on two steps. In the first one, the
stiffness and damping constants of the foundation are established. This
is achieved in the same way as in other cases involving embedded footings or pile foundations.

       In the second step the response to shock loading is analyzed. The
system usually features two masses. Because the duration of the pulse
generated by the impact of the hammer head against the anvil is very
short, typically 0.02 s, standard techniques for transient loading such
as the Fast Fourier Transform are not very suitable and special techniques are preferable. Some of them are described in Refs. 7, 11, 17, 22
and a brief review of them is given in (18). In most systems, the mass
of the anvil is too small to eliminate theoretical tension in the anvil
pad and anvil uplift (jumping) may occur which considerably affects the
whole time history of the response. The occurrence of this phenomenon,
theoretically analyzed in (11), was confirmed in a full scale experiment
conducted on the large hammer foundation depicted in Fig. 21. The
hammer features a head (ram) weighing 15,432 lb (7,000 kg) and a reinforced concrete, embedded inertial block. The anvil pad is built up of

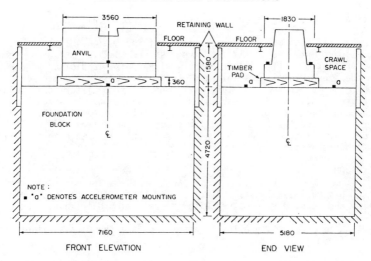

Figure 21. Outline of Large Hammer Foundation Used in Full Scale Experiment (Dimensions in mm, 1 in = 25.4 mm; Harwood and Novak, 11)

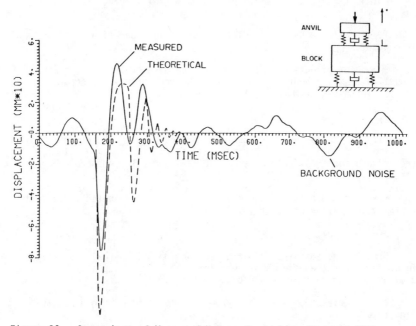

Figure 22. Comparison of Measured Hammer Foundation Response With Theoretical Prediction For the Foundation Shown in Fig. 21 (Harwood, and Novak, 11)

oak timber. The foundation is described in more detail in (11).

The stiffness and damping constants of the two mass system were evaluated as outlined in (17) and the response analyzed using the technique presented in (11). The theoretical response is plotted together with the experimentally established one in Fig. 22. The experimental response, obtained from accelerometer signals after filtering and digital processing, was affected by significant background noise generated by the operation of other facilities in the plant. Despite that, the theory predicted the fundamental frequency of the system very well and the peak amplitudes of the foundation reasonably well when anvil uplift was accounted for.

CONCLUSIONS

Much experimental data collected on vibrating foundations, both shallow and deep, were compared with theoretical predictions. The agreement between the theory and experiments was found to be, on the whole, quite encouraging but some aspects of the theories available have to be refined.

Shallow foundations:

Vertical response amplitudes tend to be overestimated which may be attributed primarily to redistribution of stresses in soil and its layering.

Torsional response is predicted poorly; incorporation of slippage is needed.

Embedment effects can be overestimated unless allowance is made for reduction of soil shear modulus towards ground surface and separation of the footing sides from the soil.

Single Piles:

Pile stiffness and damping are very sensitive to reduction of soil shear modulus towards ground surface and to pile separation. These factors are most critical for single pin-headed piles embedded in granular soils.

Fixed headed piles in cohesive soils are better suited for experimental verification of theories.

Pile Groups:

Closely spaced piles are affected considerably by pile-soil-pile interaction and the effects of this interaction on group stiffness and damping are strongly frequency dependent.

The theories available for dynamic analysis of pile groups give good results.

Static interaction coefficients applied to dynamic situations give results of limited value.

It is desirable to account for pile separation and soil nonlinearity in order to avoid overestimation of group effects.

Hammer Foundations:

The theories available predict the first amplitude of the footing and the anvil quite well but in most two mass foundations, anvil uplift can occur and its peak value may far exceed the first amplitude. This uplift should be accounted for.

REFERENCES

1. Apsel, R.J. and Luco, J.E., "Torsional response of rigid embedded foundations," J. Engrg. Mech. Div., ASCE, Vol. 102, No. EM6, 1976, pp. 957-970.

2. Awojobi, A.O. and Grootenhuis, P., "Vibrations of rigid bodies on semi-infinite elastic media," Proc. Royal Soc. of London, Series A, Vol. 287, 1965, pp. 27-63.

3. Beredugo, Y.O. and Novak, M., "Coupled horizontal and rocking vibration of embedded footings," Canadian Geotechnical Journal, Vol. 9, No. 2, 1972, pp. 477-497.

4. Bycroft, G.N., "Forced vibrations of circular plate on a semi-infinite elastic space and on an elastic stratum," Philosophical Trans. Royal Soc. Series A, Vol. 248, No. 948, 1956, pp. 327-368.

5. Constantinou, M.C. and Gazetas, G., "Torsional vibration in anisotropic halfspace," J. Geotechnical Engrg., Vol. 110, No. 11, Nov. 1984, pp. 1549-1558.

6. Dobry, R., Vicente, E., O'Rourke, M.J. and Roesset, J.M., "Horizontal stiffness and damping of single piles", J. Geotech. Engrg. Div., ASCE, Vol. 108, GT3, 1982, pp. 439-459.

7. El Hifnawy, L. and Novak, M., "Response of hammer foundations to pulse loading," Int. J. Soil Dyn. and Earthquake Engrg., Vol. 3, No. 3, 1984, pp. 124-132.

8. El Sharnouby, B. and Novak, M., "Dynamic experiments with group of piles", J. Geotech. Engrg. Div., ASCE, Vol. 110, No. 6, June 1984, pp. 719-737.

9. Fry, Z.B., "Development and evaluation of soil bearing capacity, foundations of structures; field vibratory tests data," Water-Ways Experiment Station, Technical Report No. 3-632, Report 1, 1963.

10. Gle, D.R. and Woods, R.D., "Predicted versus observed dynamic lateral response of pipe piles", Proc. of 8th World Conference on Earthquake Engrg., San Francisco, Cal., Vol. VI, July 1984, pp. 905-912.

11. Harwood, M. and Novak, M., "Uplift in hammer foundations", J. Soil Dyn. and Earthquake Engineering, Oct. 1985.

12. Kaynia, A.M. and Kausel, E., "Dynamic behavior of pile groups", 2nd Int. Conf. on Num. Methods in Offshore Piling, Austin, Texas, 1980, pp. 509-532.

13. Krishnan, R., Gazetas, G. and Velez, A., "Static and dynamic lateral deflexion of piles in non-homogeneous soil stratum", Geotechnique 33, No. 3, 1983, pp. 307-325.

14. Novak, M., "Prediction of footing vibrations", Journal of the Soil Mechanics and Foundations Div., Proc. of ASCE, Vol. 96, No. SM3, May 1970, pp. 837-861.

15. Novak, M., "Data reduction from non-linear response curves", J. Engrg. Mech. Div., ASCE, Vol. 97, No. EM4, August 1971, pp. 1187-1204.

16. Novak, M., "Vertical vibration of floating piles", J. Engrg. Mech. Div., ASCE, Vol. 103, No. EM1, Feb. 1977, pp. 153-168.

17. Novak, M., "Foundations for shock-producing machines", Can. Geotechnical Journal, Vol. 20, No. 1, 1983, pp. 141-158.

18. Novak, M., "Analysis of hammer foundations", Proc. 2nd Int. Conf. on Soil Dyn. and Earthquake Engrg., June-July 1985, QE2.

19. Novak, M., Beredugo, Y., "The effect of embedment on footing vibrations", Proc. Third Can. Conf. on Earthquake Engrg. Research, Univ. of British Columbia, Vancouver, B.C., 25-26 May, 1971, Paper No. 7, pp. 111-125.

20. Novak and Aboul-Ella, F., "Impedance functions of piles in layered media", J. Engrg. Mech. Div., ASCE, Vol. 104, No. EM3, Proc. Paper 13847, June 1978, pp. 643-661.

21. Novak, M., Aboul-Ella, F. and Sheta, M., "PILAY" and "PILAY2", A computer program for calculation of stiffness and damping of piles in layered media, SACDA, The University of Western Ontario, London, Ontario, Canada, Jan. 1981.

22. Novak, M. and El Hifnawy, L., "Vibration of hammer foundation", Int. J. Soil Dynamics and Earthquake Engrg., Vol. 2, No. 1, 1983, pp. 43-53.

23. Novak, M. and El Sharnouby, B., "Stiffness constants of single piles", J. Geotech. Engrg., ASCE, Vol. 109, No. 7, July 1983, pp. 961-974.

24. Novak, M. and El Sharnouby, B., "Evaluation of dynamic experiments on pile group", J. Geotech. Engrg. Div., ASCE, Vol. 110, No. 6, June 1984, pp. 738-756.

25. Novak, M. and Grigg, R.F., "Dynamic experiments with small pile foundations", Can. Geotechnical J., Vol. 13, No. 4, Nov. 1976, pp. 372-385.

26. Novak, M. and Howell, J.F., "Torsional vibration of pile foundations", J. Geotech. Engrg. Div., ASCE, Vol. 103, No. GT4, April 1977, pp. 271-285.

27. Novak, M. and Howell, J.F., "Dynamic response of pile foundations in torsion", J. Geotech. Engrg. Div., ASCE, Vol. 104, No. GT5, 1978, pp. 535-552.

28. Novak, M. and Sachs, K., "Torsional and coupled vibrations of embedded footings", Int. J. of Earthquake Engrg. and Struct. Dyn., Vol. 2, 11, 33, 1973.

29. Novak, M. and Sheta, M., "Approximate approach to contact problems of piles", Proc. Geotech. Engrg. Div., ASCE National Convention "Dynamic Response of Pile Foundations: Analytical Aspects", Oct. 30, 1980, pp. 53-79.

30. Novak, M. and Sheta, M., "Dynamic response of piles and pile groups", Proc. 2nd Int. Conf. on Num. Methods in Offshore Piling, Austin, Texas, April 29-30, 1982, pp. 489-507.

31. Petrovski, J., "Prediction of dynamic response of embedded foundation", Proc. 6th World Conf. on Earthquake Engrg., New Delhi, Vol. 4, 1977, pp. 169-174.

32. Poulos, H.G. and Davies, E.H., "Pile foundation analysis and design", John Wiley and Sons, 1980, pp. 397.

33. Sheta, M. and Novak, M., "Vertical vibration of pile groups", J. Geotech. Engrg. Div., ASCE, Vol. 108, No. GT4, April 1982, pp. 570-590.

34. Sung, T.Y., "Vibrations in semi-infinite solids due to periodic surface loading", ASTM Special Technical Publication, No. 156, Symposium on Dynamic Testing of Soils, 1953, pp. 35-64.

35. Tajimi, H., "Predicted and measured vibrational characteristics of a large-scale shaking table foundation", Proc. of 8th World Conf. on Earthquake Engrg., Vol. III, San Francisco 1984, pp. 873-880.

36. Veletsos, A.S. and Verbic, B., "Vibration of viscoelastic foundations", J. Earthquake Engrg. and Struct. Dyn., Vol. 2, 1973, pp. 87-102.

37. Veletsos, A.S. and Nair, V.V.D., "Torsional vibration of viscoelastic foundations", J. Geotech. Engrg. Div., ASCE, Vol. 100, GT3, March 1974, pp. 225-246.

38. Verbic, B., "Experimental and analytical analysis of soil-structure interaction", Part One: Block foundations", Institute for Materials and Structures, Faculty of Civil Engrg., Sarajevo, Feb. 1985, p. 99.

39. Waas, G. and Hartmann, H.G., "Pile foundations subjected to dynamic horizontal loads", European Simulation Meeting "Modelling and Simulation of Large Scale Structural Systems", Capri, Italy, Sept. 1981, pp. 17 (also SMIRT, Paris, 1981).

40. Weissmann, G.F., "Torsional vibrations of circular foundations", J. Soil Mech. Found. Div., Proc. ASCE, Vol. 97, No. SM9, Proc. Paper 8402, 1971, pp. 1293-1316.

VIBRATION ANALYSIS OF FOUNDATIONS ON LAYERED MEDIA

By

F. F. Tajirian[1], A.M., ASCE, and M. Tabatabaie[2], A.M., ASCE

ABSTRACT

A general procedure based on a new substructuring technique, the
flexible volume method, is applied to the solution of foundation
vibration problems. The procedure is capable of handling foundations
with complex geometries, and multiple flexible foundations of
arbitrary shape founded on the surface of, or embedded in, layered
viscoelastic soils and subjected to harmonic or transient loadings.
The methodology is implemented in a system of computer programs called
SASSI. The procedures for computing the necessary three-dimensional
impedance matrices and the overall response are summarized. To
demonstrate the effectiveness and accuracy of this method, SASSI is
used to evaluate the dynamic response (compliance functions) of a
rigid circular disk founded on the surface of a soil layer resting on
rigid base rock. The results compare favorably with more rigorous
continuum solutions. These results are also compared with the
response of circular footings on a damped halfspace to show the
effects of a fixed base on foundation response. To demonstrate the
applicability of the procedure to practical problems, results are
presented of a three-dimensional aircraft impact analysis on an
embedded tunnel.

INTRODUCTION

The design of foundations for vibrating machines and foundations
subject to other dynamic forces requires an accurate prediction of the
foundation response to these loads. A complete and rigorous analysis
must account for the following: the three-dimensional nature of the
problem, foundation flexibility, material and radiation damping of
soil, variation of soil properties with depth, embedment effects, and
interaction effects between multiple foundations through the soil.

Several substructuring methods for vibration analysis of foundations
are presented in the literature. The basic approach in all these
methods is to partition the complete soil-structure system into two
parts - - the structure and the soil. The soil medium is analyzed
first and the impedance properties (dynamic stiffnesses) at the
foundation-soil interface are established. In the second step, these

---

1. Research and Engineering Operation, Bechtel Group, Inc.,
    San Francisco, CA
2. Special Services Group, Harding Lawson Associates, Novato, CA

impedances are incorporated with the equations of motion of the
structural system, and the overall response is computed using standard
dynamic analysis procedures. For the case of foundations founded at
the surface of a homogeneous halfspace, the substructuring procedures
are simple and economical since the impedance functions can be readily
obtained directly from rigorous continuum solutions. However, in most
cases, the solution to the impedance problem is not known a priori and
must be obtained by performing a separate analysis, e.g., by the
finite element method. Furthermore, the solution of a complicated
impedance problem makes the substructuring methods less attractive as
compared to the complete finite element methods.

In this paper, a general procedure based on a new substructuring
method, the flexible volume method [Tabatabaie (6)], is applied to the
solution of foundation vibration problems. The procedure allows
vibration analysis to be performed on foundations with complex
geometries. It is possible to compute the response of multiple
flexible foundations of arbitrary shapes founded on the surface of, or
embedded in, layered viscoelastic soils and subjected to harmonic or
transient loadings. The actual foundation rigidity can also be
modeled.

The methodology differs from other substructuring techniques in the
manner in which the stiffness and the mass matrices of the foundation
are partitioned from those of the soil. The complete soil-foundation
system is divided into two substructures, the soil and the
foundation. The foundation is modeled by standard finite elements,
and the interaction is assumed to occur over the embedded volume
rather than at the boundary, i.e., at all foundation nodes below
grade. The mass and stiffness of the foundation are reduced by the
corresponding properties of the volume of soil excavated, but are
retained within the halfspace. Thus, the impedance problem is reduced
to a series of axisymmetric solutions of the response of a layered
site to point loads [Tajirian (7)]. The above methodology is
implemented in a system of interrelated computer program modules
called SASSI (5).

The flexible volume method and the procedures for computing the
necessary three-dimensional impedance matrices and the overall
response are summarized below. To demonstrate the effectiveness and
accuracy of this method, SASSI is used to evaluate the dynamic
response (compliance functions) of a rigid circular disk founded on
the surface of a soil layer resting on rigid-base rock. The results
are compared with the more rigorous continuum solutions obtained using
the computer program LUCON (4).

**METHODOLOGY FOR GENERAL THREE-DIMENSIONAL FOUNDATION VIBRATION
ANALYSIS**

Formulation of the Flexible Volume Method

The detailed formulation of the flexible volume method for analysis of
soil-structure systems is given in References (6) and (7). The
formulation below applies to forced vibration analysis of
foundations. The system is solved in the frequency domain using the

complex response method. Material damping is accounted for by using
complex material moduli. Transient loadings are decomposed by Fast
Fourier Transform techniques.

Fig. 1 represents a complete plane-strain soil-structure system
discretized by finite elements. The foundation of this system is
truncated at some far distance from the structure, and the effect of
the remaining halfspace is accounted for by introducing a set of
forces Q_b which act on the external boundary of the model. (The
selected model is chosen for explicitness only and the method is not
limited to plane-strain models nor to discretized foundations.)

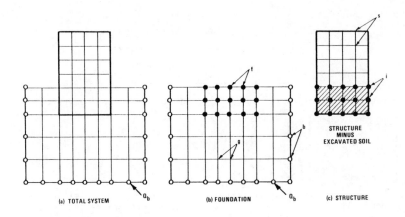

Figure 1.   Substructuring of Interaction Model

The complete soil-structure system, Fig. 1a, is partitioned into two
substructures, the foundation (Fig. 1b) and the structure (Fig. 1c).
The mass and stiffness of the structure is reduced by the
corresponding properties of the volume of soil excavated, but it is
retained within the halfspace. Furthermore, the interaction is
assumed to occur over a volume, i.e., at all basement nodes.

Using the complex response method, the discretized equation of motion
for the complete system in Figure 1a can be formulated. This equation
in the frequency domain can be written as:

$$\underline{C}^* \cdot \underline{U}^* = \underline{Q}_b^* \tag{1}$$

where $\underline{U}^*$ and $\underline{Q}_b^*$ are the vectors of complex nodal point
displacements and force amplitudes respectively and $\underline{C}^*$ is the
complex frequency- dependent stiffness matrix, which can be written as

$$\underline{C}^* = \underline{K}^* - \omega^2 \underline{M} \tag{2}$$

$\underline{M}$ and $\underline{K}^*$ are the total mass and complex stiffness matrices of the system, respectively, and are assembled using standard finite element techniques. In the above equations, the superscript "*" denotes a complex term. In the following derivation, this superscript is omitted with the understanding that all stiffness terms, displacements, and forces are complex, unless otherwise stated.

Similar equations of motion can be written for each of the substructures in Fig. 1. The following subscripts are introduced to refer to the degrees of freedom (DOF) associated with different nodes:

Subscript	Node
s	superstructure
i	basement
f	excavated soil
b	external boundary
g	remaining soil

The equation of motion for the soil (substructure 1) can be written as

$$\begin{bmatrix} \underline{C}_{ff} & \underline{C}_{fg} & \underline{C}_{fb} \\ \underline{C}_{gf} & \underline{C}_{gg} & \underline{C}_{gb} \\ \underline{C}_{bf} & \underline{C}_{bg} & \underline{C}_{bb} \end{bmatrix} \left\{ \begin{matrix} \underline{U}_f \\ \underline{U}_g \\ \underline{U}_b \end{matrix} \right\} = \left\{ \begin{matrix} \underline{Q}_f \\ 0 \\ \underline{Q}_b \end{matrix} \right\} \tag{3}$$

where $\underline{Q}_f$ are the interaction forces from the structure. Similarly, the equation of motion for the structure (substructure 2) can be written

$$\begin{bmatrix} \underline{C}_{ss} & \underline{C}_{si} \\ \underline{C}_{is} & (\underline{C}_{ii}-\underline{C}_{ff}) \end{bmatrix} \left\{ \begin{matrix} \underline{U}_s \\ \underline{U}_i \end{matrix} \right\} = \left\{ \begin{matrix} \underline{F}_s \\ \underline{F}_i+\underline{Q}_i \end{matrix} \right\} \tag{4}$$

where $\underline{F}_s$ and $\underline{F}_i$ are the amplitudes of the external forces at the superstructure and basement nodes, respectively. Compatibility of displacements and equilibrium of forces at the soil-structure interface require the following conditions:

$$\underline{U}_i = \underline{U}_f \tag{5}$$

$$\underline{Q}_i + \underline{Q}_f = \underline{0} \tag{6}$$

By substituting Eq. (6) into (4), we obtain

$$\begin{bmatrix} \underline{C}_{ss} & \underline{C}_{si} \\ \underline{C}_{is} & (\underline{C}_{ii}-\underline{C}_{ff}) \end{bmatrix} \left\{ \begin{matrix} \underline{U}_s \\ \underline{U}_i \end{matrix} \right\} = \left\{ \begin{matrix} \underline{F}_s \\ \underline{F}_i-\underline{Q}_f \end{matrix} \right\} \tag{7}$$

The term $(\underline{C}_{ii}-\underline{C}_{ff})$ simply indicates the stated partitioning according to which the stiffness and mass of the excavated soil are subtracted from the stiffness of the structure.

If any existing rock boundary is at rest and the truncated external boundary is selected infinitely far from the loaded foundation, we can assume that

$$\underline{Q}_b = \underline{0} \tag{8}$$

By substituting Eq. (8) into (3) and partitioning, we obtain:

$$\begin{bmatrix} \underline{C}_{ff} & \underline{C}_{fg} & \underline{C}_{fb} \\ \underline{C}_{gf} & \underline{C}_{gg} & \underline{C}_{gb} \\ \underline{C}_{bf} & \underline{C}_{bg} & \underline{C}_{bb} \end{bmatrix} \left\{ \begin{array}{c} \underline{U}_f \\ \underline{U}_g \\ \underline{U}_b \end{array} \right\} = \left\{ \begin{array}{c} \underline{Q}_f \\ \underline{0} \\ \underline{0} \end{array} \right\} \tag{9}$$

According to the above partitioning $\underline{U}_g$ and $\underline{U}_b$ can be eliminated and the relationship between $\underline{U}_f$ and $\underline{Q}_f$ can be expressed in the form

$$\underline{Q}_f = \underline{X}_f \cdot \underline{U}_f \tag{10}$$

$\underline{X}_f$ is a frequency-dependent matrix which represents the dynamic stiffness of the foundation soil at the interaction nodes. $\underline{X}_f$ will be referred to as the impedance matrix. An effective method for determining this matrix without using the large matrix in Eq. 9 is described in the next section.

Substitutions of Eqs. (10), (5), and (6) into Eq. (7) results in

$$\begin{bmatrix} \underline{C}_{ss} & \underline{C}_{si} \\ \underline{C}_{is} & (\underline{C}_{ii} - \underline{C}_{ff} + \underline{X}_f) \end{bmatrix} \left\{ \begin{array}{c} \underline{U}_s \\ \underline{U}_i \end{array} \right\} = \left\{ \begin{array}{c} \underline{F}_s \\ \underline{F}_i \end{array} \right\} \tag{11}$$

According to this formulation, the solution of the foundation vibration problem reduces to two steps (for each frequency):

1. Solve the impedance problem, Eq. 10, to determine the matrix $\underline{X}_f$.

2. Solve the structural problem, Eq. 11. This involves forming the complex stiffness matrices and load vector and solving Eq. 11 for the final displacements using standard equation solvers.

<u>Formulation of the Impedance Matrix</u>

According to the definition of flexibility and stiffness matrices, the impedance (dynamic stiffness) matrix, $\underline{X}_f$, for the interaction degrees of freedom can be determined as the inverse of the dynamic flexibility matrix, $\underline{F}_f$, i.e.,

$$\underline{X}_f = \underline{F}_f^{-1} \tag{12}$$

$\underline{F}_f$ is a full symmetric complex matrix. An efficient in-place inversion subroutine (8) is currently used for such operation. This method is called the "direct method" for computing the impedance matrix. Other efficient methods for computing the impedance matrix have been developed (6) and (8). These include procedures for computing the impedance matrices of symmetric and antisymmetric systems.

Formulation of the Flexibility Matrix

Procedures for determining the flexibility matrix for three-dimensional systems are described by Tajirian (7). The basic problem in determining the dynamic flexibility matrix is to find the response of a layered halfspace to a harmonic point load. Each column of the flexibility matrix is formed by applying a unit point load at the interaction degree-of-feeedom associated with that column and by computing the resulting displacements at all the interaction nodes.

For layered sites, these displacements can be obtained from the axisymmetric model shown in Fig. 2. This model consists of a central core of special cylindrical axisymmetric finite elements connected at the perimeter to a semi-infinite layered zone which is represented by axisymmetric transmitting boundaries (3) and (10). Either the lower boundary can be fixed or a halfspace can be simulated by using the variable depth and viscous boundary methods (1), (6) and (7).

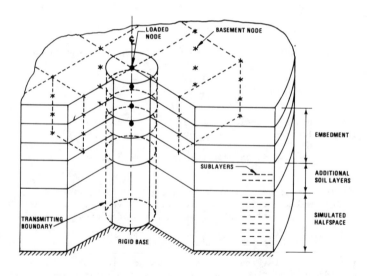

Figure 2.   Axisymmetric Model for Impedance Analysis

From this model displacement amplitudes can be obtained both at the central nodes and at any point outside the cylindrical elements, i.e., at all interaction nodes. These displacements which are computed in a cylindrical coordinate system are transformed to the global cartesian coordinate system using standard transformation procedures.

CASE STUDIES

Case Study 1:   Compliance Functions For Circular Disk
                on a Layer Overlying a Rigid Base

In this section, the accuracy of the above methodology is tested by
comparing SASSI results with known solutions for the problem of forced
vibration of rigid foundations on viscoelastic layered media.

Compliance functions for a rigid circular disk on a uniform layer of
finite thickness overlying a horizontal rigid base were obtained using
the computer program LUCON (4). This program uses a special
formulation of the field problem in terms of Green's functions and can
handle a layered viscoelastic halfspace.

The same problem was solved using SASSI. Fig. 3 shows the geometry,
the material properties, and the finite element mesh used in the SASSI
analysis. Because the results of the analysis are presented in terms
of the dimensionless frequency ratio $A_O = \omega r/V_S$, the parameters of
the problem can be selected arbitrarily as long as the frequencies of
analysis, ω , are adjusted accordingly to cover the presented range
of A_O.

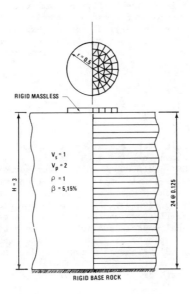

Figure 3.   SASSI Model for Computing Compliance Functions

The compliance functions for 5 and 15 percent layer damping obtained using the SASSI procedure are compared with comparable LUCON results. These results are also presented along with damped halfspace solutions (9) to investigate the effectiveness of the model consisting of a finite layer on a rigid base rock (Fig. 3) in simulating halfspace conditions. Figs. 4 and 5 show the vertical compliance functions for 5% and 15% damping, respectively. Excellent agreement exists between SASSI and LUCON results except for the real part of the 5% compliance at A_O values of less than 0.2. Considerable deviation of the results from halfspace solution can be seen especially near the peaks corresponding to vertical natural frequencies of the system. These frequencies can be obtained from

$$F_{vnf} = (2n-1) V_p/(4H) \quad , n=1,2,3...$$

For the present problem ($V_p = 2$, $H = 3$), $F_{vnf} = 0.166, 0.500, 0.833, ...$
and thus: $A_{vnf} = 0.524, 1.571, 2.618, ...$

However, the results seem to be less sensitive to occurrence of the peaks at higher modes especially when the damping is high. In general, the vertical compliance of the model shown in Fig. 3 poorly represents the damped halfspace results at A_O less than 1.5.

Comparisons of horizontal compliance functions are shown in Figs. 6 and 7. Both methods yield essentially the same results. The rigid-base results tend to better match the halfspace results for the horizontal case than for the vertical case. This is in part due to the fact that the peaks corresponding to the horizontal natural frequencies of the layer on rigid base start from much lower frequencies.

$$F_{hnf} = (2n-1) V_s/(4H) \quad n=1,2,3. \ ...$$

thus for $V_s = 1$, $H = 3$, we obtain $F_{hnf} = 0.083, 0.250, 0.417, ...$
and $A_{hnf} = 0.262, 0.785, 1.309, ...$

Comparison of the results for rocking and torsional cases for 5 and 15 percent damping, as shown in Figs. 8 through 11, indicate very good agreement between SASSI and LUCON results. Furthermore, it appears that the rocking and torsional compliance functions obtained from the model shown in Fig. 3 can be used to represent those of the damped halfspace case with good accuracy. This is in agreement with the findings of other investigators who have shown that the compliance functions for rotational sources are not influenced by the presence of a rigid base when the depth to the base is greater than three times the foundation radius (3). This is due to the destructive interference of waves emanating from the footing-soil interface which limit the effective depth of penetration of the generated waves.

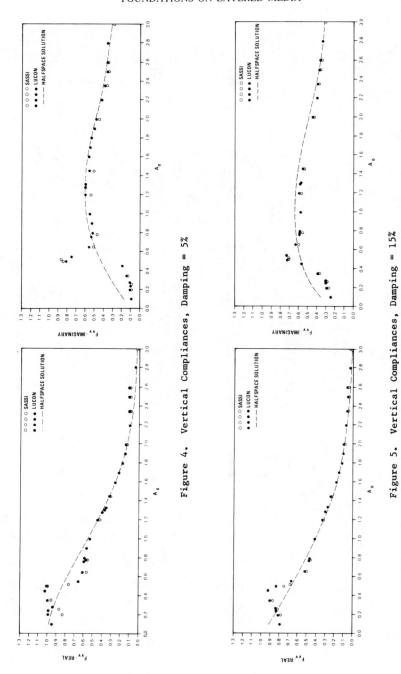

Figure 4. Vertical Compliances, Damping = 5%

Figure 5. Vertical Compliances, Damping = 15%

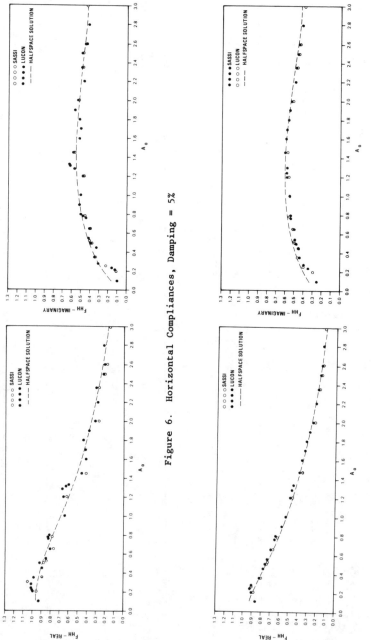

Figure 6. Horizontal Compliances, Damping = 5%

Figure 7. Horizontal Compliances, Damping = 15%

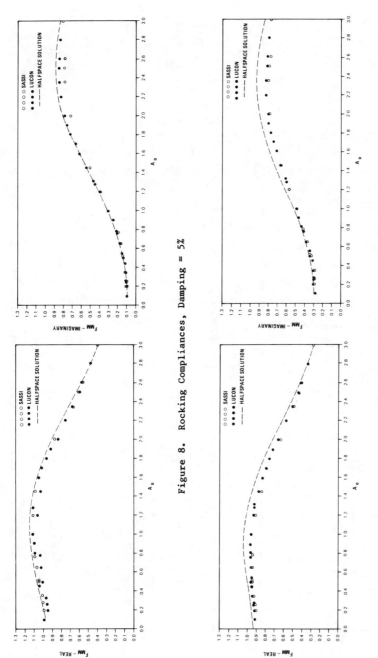

Figure 8. Rocking Compliances, Damping = 5%

Figure 9. Rocking Compliances, Damping = 15%

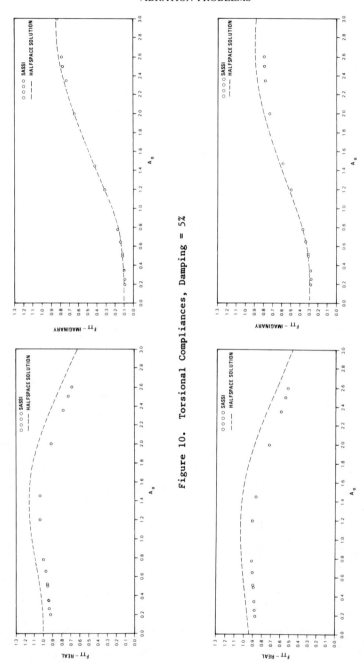

Figure 10. Torsional Compliances, Damping = 5%

Figure 11. Torsional Compliances, Damping = 15%

Case Study 2: <u>Three-Dimensional Analysis of Airplane Impact on an</u>
<u>Underground Cable Tunnel and Protective Slab</u>

The second case study is a three-dimensional analysis of an underground
cable tunnel. The response of the tunnel to an aircraft impact on a
protective slab at grade is computed.

The tunnel which runs between a reactor and control building at a nuclear
power plant is shown in Figs. 12 and 13. The tunnel has a height of
about 15 ft and is 35.8 ft wide. It is designed with a protective slab
about 3 ft thick at the ground level and with an intermediate layer of
earth about 3.25 ft thick. To prevent impact on the sides of the tunnel,
the protective slab is wider than the tunnel. The thickness of the
protective slab is designed to prevent perforation in the direct loading
area where an airplane might impact. Thus, the protective slab and the
tunnel have to be designed to withstand stresses during the
time-dependent loading shown in Fig. 14.

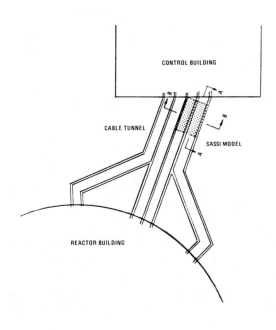

Figure 12. Plan View of Tunnel

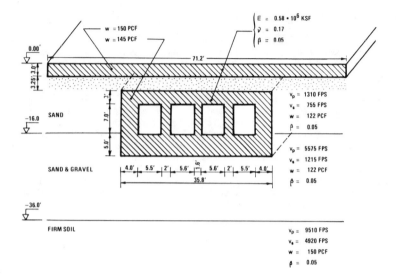

Figure 13.   Section Through Tunnel

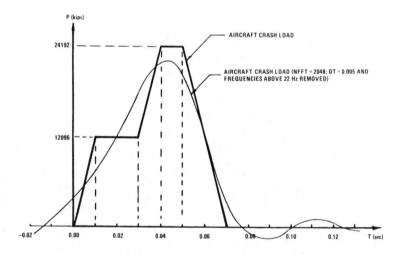

Figure 14.   Input Time History

The soils around the tunnel consist of about 16 ft of sand underlain successively by 20 ft of sand and gravel and a deep bed of firm soils (Fig. 13).

A standard German aircraft crash loading was used as input. The loading was assumed to act vertically at the center of the protective slab and was distributed over an area of 75 ft^2. The maximum amplitude of the load was 24,192 kips. In the analysis, one quarter of the total load was distributed on the symmetric model of the prototype slab (Fig. 15).

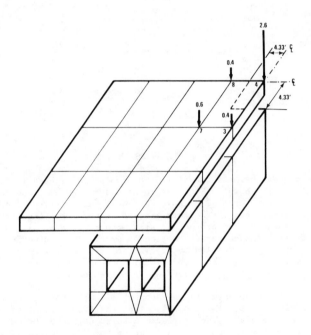

Figure 15.   Input Load Distribution

The first 0.16 seconds of the load digitized at 0.005 seconds was used. The record consisted of 2048 points. The maximum frequency considered was 22 Hz. The filtered time history is compared with the original loading (Fig. 14).

The soil profile used in the SASSI analysis is shown in Fig. 16. Layer thicknesses were chosen to be less than one fifth of the shear wave length at the cutoff frequency of 22 Hz; i.e., allowable thickness $V_s/(5*22)$. The underlying halfspace (firmsoil) was simulated using 20 sublayers whose thicknesses vary with frequency attached to viscous dashpots.

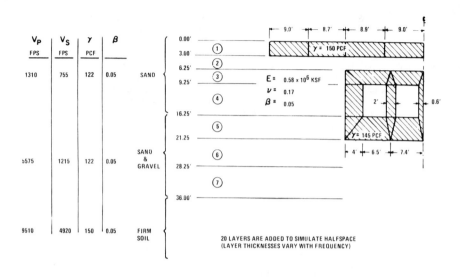

Figure 16.   SASSI Model

The three-dimensional structural finite element model used in the
SASSI analysis is shown in Fig. 17. Since advantage could be taken of
symmetry about two vertical planes, only one quarter of the tunnel
section was analyzed. Furthermore, it was assumed that the effects of
impact on the tunnel and protective slab were minor at a distance of
about 24 ft from the center of the loaded area. The protective slab
and tunnel walls were modeled by special solid-brick elements which
behave well in bending. The total model consists of 104 nodes, 45
structural elements, and 51 excavated soil elements.

Maximum vertical displacements were computed in the tunnel and the
protective slab. The largest displacement was 0.63 in., at the center
of the protective slab. The displacements in the tunnel below the
loaded area did not exceed 0.07 in. Furthermore, at points away from
the center of the loaded area the effect of impact was significantly
reduced.

Vertical time histories at the center of the protective slab (node 4)
and at the bottom of the tunnel (node 96) were compared. Such
comparisons for displacements are shown in Fig. 18, and for
accelerations in Fig. 19. As may be seen from these results the
protective slab significantly reduced the motions in the tunnel.

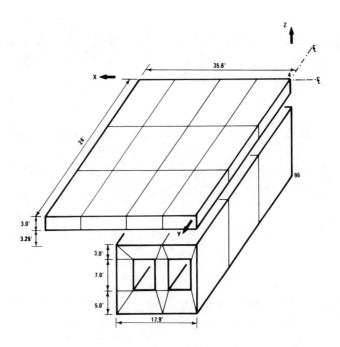

Figure 17.   Three-Dimensional Finite Element Model

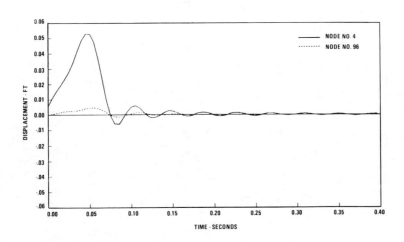

Figure 18.   Comparison of Vertical Displacement Time History

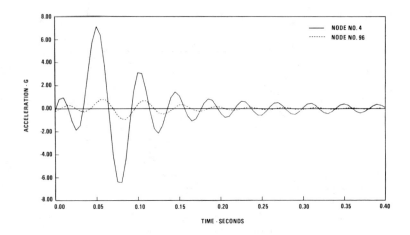

Figure 19. Comparison of Vertical Acceleration Time History

Vertical acceleration response spectra are shown in Fig. 20 (Node 4) and Fig. 21 (Node 96). The results indicate a peak in the response around the cutoff frequency (22 Hz). Thus, it may be necessary to repeat the entire analysis using a finer model which satisfied a higher frequency cutoff.

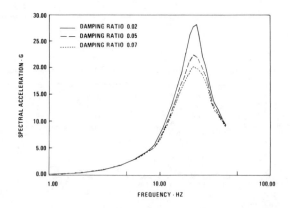

Figure 20. Absolute Vertical Acceleration Response Spectrum at Node 4

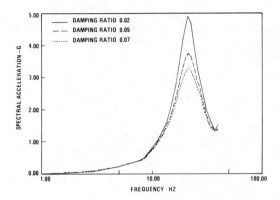

Figure 21. Absolute Vertical Acceleration Response Spectrum at Node 96

Comparisons of maximum velocities and accelerations at the same nodes are shown below. Again, the protective slab significantly reduced the motions in the tunnel.

	Node 4	Node 96
Maximum vertical displacement (ft)	0.053	0.004
Maximum vertical velocity (ft/sec)	2.80	0.33
Maximum vertical acceleration (g)	7.2	0.95

SUMMARY AND CONCLUSIONS

The flexible volume method was applied to the solution of general three-dimensional foundation vibration problems. The formulation can handle multiple flexible foundations with arbitrary shapes founded on the surface of, or embedded in, layered viscoelastic soils. Responses computed using this method compared favorably with more rigorous continuum solutions, demonstrating its accuracy. To illustrate the applicability of the procedure to more practical problems, the results of a three-dimensional analysis of an aircraft impact on a buried tunnel were presented.

ACKNOWLEDGMENTS

The flexible volume method and the computer Program SASSI were developed at the University of California, Berkeley, under the supervision of Professor John Lysmer. The authors gratefully acknowledge his contributions and support. The airplane impact study presented in this paper was sponsored by Kraftwerk Union AG, Germany. This support is also acknowledged with appreciation. Preparation of this manuscript was assisted by Bechtel Group Inc. These contributions are acknowledged with appreciation.

APPENDIX. – REFERENCES

1. Chen, J.-C., "Analysis of Local Variations in Free Field Seismic Ground Motions," thesis presented to the University of California, at Berkeley, Calif., in 1980, in partial fulfillment for the requirements of the degree of Doctor of Philosophy.

2. Gazetas, G., "Analysis of Maching Foundation Vibrations: State of the Art," Soil Dynamics and Earthquake Engineering, Vol. 2., pp 2-42, 1983.

3. Kausel, E., "Forced Vibrations of Circular Foundations on Layered Media," Soils Publication No. 336, Dept. of Civil Engineering, Massachusetts Institute of Technology, Cambridge, Mass., January 1974.

4. Luco, J. E. "LUCON – Impedance Functions for a Rigid Circular Foundation on a Layered Viscoelastic Medium," Bechtel Power Corporation, November 1980.

5. Lysmer, J., Tabatabaie, M., Tajirian, F., Vahdani, S. and Ostadan, F., "SASSI – A System for Analysis of Soil-Structure Interaction," Report No. UCB/GT/81-02, Geotechnical Engineering, Dept. of Civil Engineering, University of California, Berkeley, April 1981.

6. Tabatabaie, M., "The Flexible-Volume Method for Dynamic Soil-Structure Interaction Analysis", thesis presented to the University of California, at Berkeley, Calif., in 1982, in partial fulfillment of the requirements for the degree of Doctor of Philosophy.

7. Tajirian, F., "Impedance Matrices and Interpolation Techniques for 3-D Interaction Analysis by the Flexible Volume Method," thesis presented to the University of California, at Berkeley, Calif., in 1981, in partial fulfillment of the requirements for the degree of Doctor of Philosophy.

8. Vahdani, S., "Impedance Matrices for Soil-Structure Interaction Analysis by the Flexible Volume Method," thesis presented to the University of California, at Berkeley, Calif., in 1983, in partial fulfillment of the requirements for the degree of Doctor of Philosophy.

9. Veletsos, A. S. and Verbic, B., "Vibration of Viscoelastic Foundation," Earthquake Engineering and Structural Dynamics, Vol. 2, pp. 87-102, 1973.

10. Waas, G., "Analysis Method for Footing Vibration Through Layered Media," thesis presented to the University of California, at Berkeley, Calif., in 1972, in partial fulfillment of the requirements for the degree of Doctor of Philosophy.

SIMPLE APPROACH FOR EVALUATION OF COMPLIANCE MATRIX OF PILE GROUPS

Toyoaki Nogami,* M. ASCE and Kazuo Konagai**

A simple and efficient method is presented for the computation of the compliance matrix of pile groups subjected to vibrations at the heads. The method requires numerical analyses only for single piles attached to a Winkler soil model and can be applied to any complex soil profiles. Charts are also presented for quick evaluation of the compliance matrix of pile groups in a homogeneous soil profile. The frequency range considered includes high frequencies, which are often encountered in machine induced vibrations.

Introduction

Piles are often used in a group as a part of a foundation system. When a machine foundation system contains a group of piles, the vibrations induced by a machine are transmitted to the piles and then to the soil medium through the pile shafts. This results in a pile-soil-pile interaction among the piles. It has been found that such interaction effects (group effects) are more significant under dynamic than under static loads and can alter the dynamic response of the system significantly (e.g., 1,2,3,4,6,9,10,11,12). Thus, the group effects must be considered in a rational manner in the analysis of pile groups subjected to dynamic loads.

Various methods have been developed to analyze the dynamic responses of pile groups taking a pile-soil-pile interaction into account (e.g., 1,3,4,12). They usually require a large effort in a numerical analysis. Nogami (6) has recently developed a simple approach and prepared useful charts for the evaluation of pile group flexibility (compliance) matrix in flexural vibration caused by horizontal loads and rocking moments. However, they are limited to the frequency range relevant to seismic loading and thus does not include the high frequency range often encountered in machine induced vibrations.

Construction of Compliance Matrix of Pile Group

A flexibility matrix (compliance matrix) of an elastic pile group is often obtained from those of all sets of two piles in the entire

---

*Assoc. Prof. of Civil Engineering, University of Houston - University Park, Houston, TX 77004.

**Assoc. Prof. of Civil Engineering, Technological University of Nagaoka, Nagaoka, Japan.

group(5,7,8). This approximate approach is well accepted for the static problems, and has been recently confirmed to be applicable also to the dynamic problems (1,9,10). This approximation can reduce the computational effort in evaluating the compliance matrix of a pile group, and thus is adopted in the present approach.

Fig. 1 shows the ith and j^{th} piles in the group and the horizontal loval coordinates u and v. The local coordinate w is perpendicular to the plane shown in the figure. By considering only the ith and j^{th} piles in the group, the pile-head responses of the ith pile can be expressed as

$$\left\{ \begin{array}{c} u_i^{u,v} \\ \psi_i^{v,u} \end{array} \right\} = \left[\; F_{ii}^{u,v} \; \right] \left\{ \begin{array}{c} P_i^{u,v} \\ M_i^{v,u} \end{array} \right\} + \left[\; F_{ij}^{u,v} \; \right] \left\{ \begin{array}{c} P_j^{u,v} \\ M_j^{v,u} \end{array} \right\}$$

$$u_i^w = F_{ii}^w \; P_i^w + F_{ij}^w \; P_J^w \tag{1}$$

where u^u, u^v, and u^w = amplitudes of the translational pile-head displacements in u, v and w directions, respectively; ψ^u and ψ^v = amplitudes of rotational pile-head displacements around u and v axis, respectively; and P^u, P^v, and P^w = amplitudes of the forces applied in the u, v, and w directions, respectively; M^v and M^u = amplitudes of the moment around v and u axes, respectively. $F^{u,v}$ is the flexural compliance of the pile head, and F^w is the vertical compliance of the pile head. The first and second terms of the right-hand side of Eq. 1 corresponds to the displacements of the ith pile due to the loads applied at the ith and j^{th} piles, respectively.

Ignoring the piles other than the ith and j^{th} piles, the responses of the ith and j^{th} piles subjected to identical loads at both pile heads can be expressed using Eq. 1 as

$$\left\{ \begin{array}{c} u^{u,v} \\ \psi^{v,u} \end{array} \right\} = \left[\; F_*^{u,v} \; \right] \left\{ \begin{array}{c} P^{u,v} \\ M^{v,u} \end{array} \right\}$$

$$u_i^w = F_*^w \; P^w \tag{2}$$

where

$$[F_*^{u,v}] = [F_{ii}^{u,v}] + [F_{ij}^{u,v}]$$

$$F_*^w = F_{ii}^w + F_{ij}^w \tag{3}$$

Since F_{ii} is approximately equal to that for a single pile, F_o, the expression for F_{ij} can be obtained from Eq. 3 as

$$[F_{ij}^{u,v}] = [F_{*}^{u,v}] - [F_{o}^{u,v}]$$

$$F_{ij}^{w} = F_{*}^{w} - F_{o}^{w} \qquad (4)$$

where $[F_{o}^{u,v}]$ and F_{o}^{w} = compliances of a single pile in a flexural and axial vibrations, respectively.

The compliance matrix of the pile group in global coordinate system is defined here as

$$\left\{ \begin{array}{c} \left\{ \begin{array}{c} \tilde{x}^{x} \\ \tilde{\psi}^{y} \end{array} \right\} \\ \left\{ \begin{array}{c} \tilde{x}^{y} \\ \tilde{\psi}^{x} \end{array} \right\} \end{array} \right\} = \begin{bmatrix} [F^{xx}] & [F^{xy}] \\ \\ [F^{yx}] & [F^{yy}] \end{bmatrix} \left\{ \begin{array}{c} \left\{ \begin{array}{c} \tilde{p}^{x} \\ \tilde{y} \\ M \end{array} \right\} \\ \left\{ \begin{array}{c} \tilde{p}^{y} \\ \tilde{x} \\ M \end{array} \right\} \end{array} \right\}$$

$$\{x^{z}\} = [F^{z}] \{P^{z}\} \qquad (5)$$

The compliance matrix in this global system can be constructed from F_{ij} and F_{o} through the following relationship between the global and local coordinate systems (see Fig. 1 for definition of θ_{ij}):

$$\left\{ \begin{array}{c} x \\ y \\ z \end{array} \right\} = \begin{bmatrix} \cos \theta_{ij} & -\sin \theta_{ij} & 0 \\ \sin \theta_{ij} & \cos \theta_{ij} & 0 \\ 0 & 0 & 1 \end{bmatrix} \left\{ \begin{array}{c} u \\ v \\ w \end{array} \right\} \qquad (6)$$

For example, the values in the compliance matrices related to the ith and j^{th} piles are

$$\left\{ \begin{array}{c} F^{xx}(i,j) \\ F^{xx}(i,N+j) \\ F^{xx}(N+i,j) \\ F^{xx}(N+i,N+j) \end{array} \right\} = \cos^{2}\theta_{ij} \left\{ \begin{array}{c} F_{ij}^{u}(u,P) \\ F_{ij}^{u}(u,M) \\ F_{ij}^{u}(\psi,P) \\ F_{ij}^{u}(\psi,M) \end{array} \right\} + \sin^{2}\theta_{ij} \left\{ \begin{array}{c} F_{ij}^{v}(u,P) \\ F_{ij}^{v}(u,M) \\ F_{ij}^{v}(\psi,P) \\ F_{ij}^{v}(\psi,M) \end{array} \right\}$$

$$
\left.\begin{matrix} F^{xy}(i,j) \\ F^{xy}(i,N+j) \\ F^{xy}(N+i,j) \\ F^{xy}(N+i,N+j) \end{matrix}\right\} = \left.\begin{matrix} F^{yx}(i,j) \\ F^{yx}(i,N+j) \\ F^{yx}(N+i,j) \\ F^{yx}(N+i,N+j) \end{matrix}\right\} = \cos\theta_{ij}\sin\theta_{ij} \left.\begin{matrix} F^{u}_{ij}(u,P) \\ F^{u}_{ij}(u,M) \\ F^{u}_{ij}(\psi,P) \\ F^{u}_{ij}(\psi,M) \end{matrix}\right\}
$$

$$
- \cos\theta_{ij}\sin\theta_{ij} \left.\begin{matrix} F^{v}_{ij}(u,P) \\ F^{v}_{ij}(u,M) \\ F^{v}_{ij}(\psi,P) \\ F^{v}_{ij}(\psi,M) \end{matrix}\right\}
$$

$$
\left.\begin{matrix} F^{yy}(i,j) \\ F^{yy}(i,N+j) \\ F^{yy}(N+i,j) \\ F^{yy}(N+i,N+j) \end{matrix}\right\} = \sin^2\theta_{ij} \left.\begin{matrix} F^{u}_{ij}(u,P) \\ F^{u}_{ij}(u,M) \\ F^{u}_{ij}(\psi,P) \\ F^{u}_{ij}(\psi,M) \end{matrix}\right\} + \cos^2\theta_{ij} \left.\begin{matrix} F^{v}_{ij}(u,P) \\ F^{v}_{ij}(u,M) \\ F^{v}_{ij}(\psi,P) \\ F^{v}_{ij}(\psi,M) \end{matrix}\right\}
$$

$$
F^{z}(i,j) = F^{w}_{ij}
$$

and

$$
\left.\begin{matrix} F^{xx}(i,i) \\ F^{xx}(i,N+i) \\ F^{xx}(N+i,i) \\ F^{xx}(N+i,N+i) \end{matrix}\right\} = \left.\begin{matrix} F^{yy}(i,i) \\ F^{yy}(i,N+i) \\ F^{yy}(N+i,i) \\ F^{yy}(N+i,N+i) \end{matrix}\right\} = \left.\begin{matrix} F_{o}(u,P) \\ F_{o}(u,M) \\ F_{o}(\psi,P) \\ F_{o}(\psi,M) \end{matrix}\right\}
\tag{7}
$$

$$
\left.\begin{matrix} F^{xy}(i,i) \\ F^{xy}(i,N+i) \\ F^{xy}(N+i,i) \\ F^{xy}(N+i,N+i) \end{matrix}\right\} = \left.\begin{matrix} F^{yx}(i,i) \\ F^{yx}(i,N+i) \\ F^{yx}(N+i,i) \\ F^{yx}(N+i,N+i) \end{matrix}\right\} = \left.\begin{matrix} 0 \\ 0 \\ 0 \\ 0 \end{matrix}\right\}
$$

where

$$[F_{ij}^{u,v}] = \begin{bmatrix} F_{ij}^{u,v}(u,P) & F_{ij}^{u,v}(u,M) \\ F_{ij}^{u,v}(\psi,P) & F_{ij}^{u,v}(\psi,M) \end{bmatrix} \tag{8}$$

$$[F_o] = \begin{bmatrix} F_o(u,P) & F_o(u,M) \\ F_o(\psi,P) & F_o(\psi,M) \end{bmatrix}$$

Computation of F_* and F_o

Soil Reactions. — Two identical piles are assumed to be interconnected by horizontal springs, "Winkler model for a pile group" (2,3,4,6), as shown in Fig. 2. When identical dynamic loads are applied at the heads of both piles, the soil displacements and the soil reactions along the pile shaft at any given depth are identical for the two piles and can be expressed as

$$u^{u,v,w} = f_o^{u,v,w} \ p^{u,v,w} + \Delta f^{u,v,w} \ p^{u,v,w} \tag{9}$$

where u and p = amplitudes of the soil displacement and soil reaction force at the given depth, respectively; and the first and second terms of the right-hand side of Eq. 9 correspond respectively to the soil displacements for a single isolated pile and additional soil displacement due to the second pile. Thus, the soil reaction to each pile is expressed as

$$p^{u,v,w} = k_*^{u,v,w} \ u^{u,v,w} \tag{10}$$

where k_* is the soil stiffness defined as

$$k_*^{u,v,w} = \frac{1}{f_o^{u,v,w} + \Delta f^{u,v,w}} \tag{11}$$

The soil stiffness for a single pile, k_o, is

$$k_o^{u,v,w} = \frac{1}{f_o^{u,v,w}} \tag{12}$$

The rigorous expressions for $f_o^{u,v,w}$ and $\Delta f^{u,v,w}$ can be found in Refs. 3, 4 and 6. However, the computations involved in those expressions are tedious, and thus the following approximate expressions may be used instead:

$$f_o^{u,v} = \frac{1}{2\pi G} \left\{ \frac{\delta_1(\nu)}{g_1(a_o)} + \delta_2(\nu) \cdot a_o^2 \right\}^{-1}$$

$$f^w_o = \frac{1}{2\pi G} \ g_1(a_o)$$

(13)

$$\Delta f^u = \frac{1}{2\pi G} \ [\ \frac{e^{i\phi a_o}}{\eta^2} \ \{g_2(b_o, \ \overline{R}) + g_3(b_o, \ \overline{R})\} - e^{-i\phi a_o} \ g_3(a_o, \ \overline{R})]$$

$$\Delta f^v = \frac{1}{2\pi G} \ [\ - \frac{e^{i\phi a_o}}{\eta^2} \ g_3(b_o, \ \overline{R}) + e^{-i\phi a_o} \{g_2(a_o, \overline{R}) + g_3(a_o, \overline{R})\}]$$

$$\Delta f^w = \frac{1}{2\pi G} \ g_2(a_o, \ \overline{R})$$

where v_s = shear wave velocity of soil at the given depth; $\delta_1(v)$ and $\delta_2(v)$ = Poisson's ratio dependent constants given in Table 1; $a_o = r_o \ \omega/v_s$; $\eta = \sqrt{2(1-v)/(1-2 \ v)}$; $b_o = a_o/\eta$; $\phi = 0.25 \times (\eta-1)/\eta$; v and G = Poisson's ration and shear modulus of soil, respectively; and $\overline{R} = R/r_o$ in which R is the distance between the two piles. The functions g_1, g_2 and g_3 in Eq. 13 are defined as

$$g_1(a_o) = \{- a^2_o + i \ a_o + 0.065\}^{-0.5}$$

$$g_2(a_o,\overline{R}) = \{- 0.757 \ \overline{R} \ a^2_o + i \ 0.85 \ \overline{R} \ a_o + 0.065\}^{-0.5} \ e^{-ia_o(\overline{R}-0.87)}$$

$$g_3(a_o, \ \overline{R} \ ) = [-(a_o\overline{R})^{-2} - i \ (a_o\overline{R})^{-1} - \{-(\overline{R} + 3.5)^2 \ a^2_o +$$

$$i2C(\overline{R}) \ a \ \}^{-0.5}] \ e^{-i \ a_o(\overline{R}-0.5)}$$

(14)

where $C(\overline{R}) = (1.082 \ \ell n \ \overline{R}-0.8414)(\overline{R} + 3.5)$ for $\overline{R} <20$ and $C(\overline{R})=2.5 \times (\overline{R}+3.5)$ for $\overline{R} \geq 20$. When the material damping in the soil is considered, a_o, b_o, v_s, and G are computed using a complex shear modulus defined as $G(1+2Di)$ in which D is the material damping parameter. The above approximate expressions can provide values very close to those computed by the rigorous expressions as shown in Fig. 3.

Pile Responses. - The pile-head responses F_* and F_o can be computed by analyzing a single pile attached to a Winkler model with spring stiffness k_* and k_o, respectively. The equations of motion of a single pile attached to the Winkler model are

$$EI \ \frac{d^4 u^{,v}}{dz^4} - m\omega^2 \ u^{u,v} = -k^{u,v} \ u^{u,v}$$

(15)

Table 1

Functions $\delta_1(\upsilon)$ and $\delta_2(\upsilon)$

Poisson's Ratio υ	$\delta_1(\upsilon)$	$\delta_2(\upsilon)$
0.50	2.000	0.5000
0.49	1.940	0.3914
0.48	1.883	0.3210
0.47	1.831	0.2668
0.46	1.784	0.2232
0.45	1.741	0.1870
0.43	1.667	0.1314
0.40	1.580	0.0714
0.35	1.476	0.0176
0.25	1.351	0
0.20	1.311	0
0.10	1.252	0
0.00	1.213	0

Table 2

Amplitudes and Phase Shifts of $F^{u,v,w}$ for Piles 1 and 10
in Example Pile Group in Homogeneous Soil

	Amplitude Ratios	Phase Shifts (radian)
$F^u(u,P)$	0.178	-3.436
$F^u(u,M)$	0.119	-3.255
$F^u(\psi,P)$	0.119	-3.255
$F^u(\psi,M)$	0.060	-3.073
$F^v(u,P)$	0.238	-5.337
$F^v(u,M)$	0.157	-5.157
$F^v(\psi,P)$	0.157	-5.157
$F^v(\psi,M)$	0.078	-4.976
F^w	0.137	-4.990

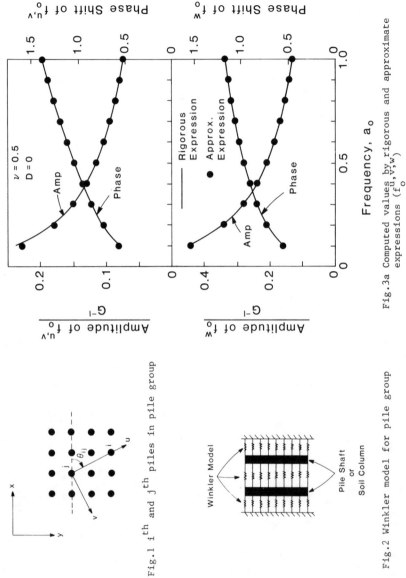

Fig.1 i^{th} and j^{th} piles in pile group

Fig.2 Winkler model for pile group

Fig.3a Computed values by rigorous and approximate expressions ($f_o^{u,v,w}$)

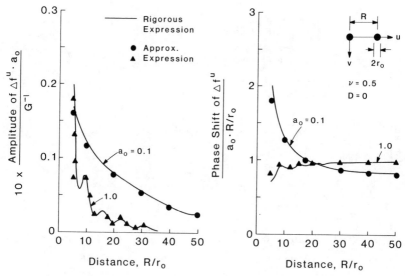

Fig.3b Computed values by rigorous and approximate expressions (Δf^u)

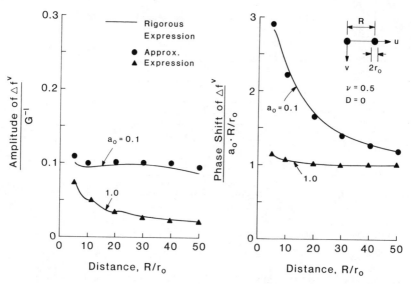

Fig.3c Computed values by rigorous and approximate expressions (Δf^v)

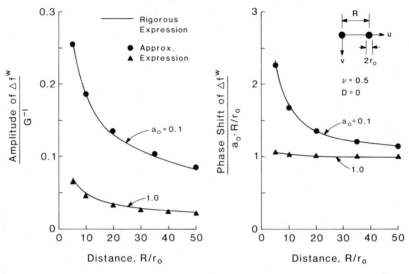

Fig.3d Computed values by rigorous and approximate expressions (Δf^w)

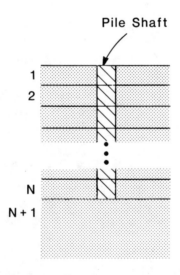

Fig.4 Soil-pile system divided into horizontal slices

$$-EA \frac{d^2 u^w}{dz^2} - m\omega^2 u^w = -k^w u^w$$

where k = either k_o or k_x; EI and EA = flexural and axial stiffnesses of the pile, respectively; ω = circular excitation frequency; and m = mass per unit length of the pile. General expressions for pile response can be obtained by solving Eq. 15 for $u^{u,v,w}$.

A soil-pile system is divided into a number of horizontal slices as shown in Fig. 4. Each of the slices contain a homogenous soil layer and pile segment. Considering a compatibility and equilibrium conditions between the ith and $i+1^{th}$ segments, the following expressions can be obtained from a general solution for Eq. 15.:

$$\left\{ \begin{matrix} u^{u,v} \\ \psi^{v,u} \end{matrix} \right\}_i = [t_{11,i}^{u,v}] \left\{ \begin{matrix} u^{u,v} \\ \psi^{v,u} \end{matrix} \right\}_{i+1} + [t_{12,i}^{u,v}] \left\{ \begin{matrix} P^{u,v} \\ M^{v,u} \end{matrix} \right\}_{i+1}$$

$$\left\{ \begin{matrix} P^{u,v} \\ M^{v,u} \end{matrix} \right\}_i = [t_{21,i}^{u,v}] \left\{ \begin{matrix} u^{u,v} \\ \psi^{v,u} \end{matrix} \right\}_{i+1} + [t_{22,i}^{u,v}] \left\{ \begin{matrix} P^{u,v} \\ M^{v,u} \end{matrix} \right\}_{i+1}$$

$$\left\{ \begin{matrix} u^w \\ P^w \end{matrix} \right\}_i = [t_i^w] \left\{ \begin{matrix} u^w \\ P^w \end{matrix} \right\}_{i+1} \tag{16}$$

where $(u \; \psi)^T$ and $(P \; M)^T$ = the displacement and force vectors at the upper end of the ith segment; $[t_i]$ = 2x2 matrix for the ith segment given in Appendix I.

The pile-tip soil stiffnesses, $[k_b^{u,v}]$ and k_b^w, are defined as

$$\left\{ \begin{matrix} P^{u,v} \\ M^{v,u} \end{matrix} \right\} = [k_b^{u,v}] \left\{ \begin{matrix} u^{u,v} \\ \psi^{v,u} \end{matrix} \right\}$$

$$P^w = k_b^w \; u^w \tag{17}$$

When the soil stiffness at the pile-tip is known, Eq. 16 for $i = N$ can be written as

$$\left\{ \begin{matrix} u^{u,v} \\ \psi^{v,u} \end{matrix} \right\}_N = [S_N^{u,v}] \left\{ \begin{matrix} u^{u,v} \\ \psi^{v,u} \end{matrix} \right\}_{N+1}$$

$$\left\{ \begin{matrix} P^{u,v} \\ M^{v,u} \end{matrix} \right\}_N = [T_N^{u,v}] \left\{ \begin{matrix} u^{u,v} \\ \psi^{v,u} \end{matrix} \right\}_{N+1} \tag{18}$$

$$\left\{ \begin{matrix} u^w \\ p^w \end{matrix} \right\}_N = \left\{ Q^w_N \right\} u^w_{N+1}$$

where $[S^{u,v}_N] = [t^{u,v}_{11,N}] + [t^{u,v}_{12,N}] [k^{u,v}_b]$

$$[T^{u,v}_N] = [t^{u,v}_{21,N}] + [t^{u,v}_{22,N}] [k^{u,v}_b]$$

(19)

$$\left\{ Q^w_N \right\} = [t^w_N] \left\{ \begin{matrix} 1 \\ k^w_b \end{matrix} \right\}$$

Successive application of Eq. 16 starting with Eq. 18 leads to

$$\left\{ \begin{matrix} u^{u,v} \\ \psi^{v,u} \end{matrix} \right\}_i = [S^{u,v}_i] \left\{ \begin{matrix} u^{u,v} \\ \psi^{v,k} \end{matrix} \right\}_{N+1}$$

$$\left\{ \begin{matrix} p^{u,v} \\ M^{v,u} \end{matrix} \right\}_i = [T^{u,v}_i] \left\{ \begin{matrix} u^{u,v} \\ \psi^{v,u} \end{matrix} \right\}_{N+1}$$

(20)

$$\left\{ \begin{matrix} u^w \\ p^w \end{matrix} \right\}_i = \left\{ Q^w_i \right\} u^w_{N+1}$$

where

$$[\hat{S}^{u,v}_i] = [t^{u,v}_{11,i}] [S^{u,v}_{i+1}] + [t^{u,v}_{12,i}] [T^{u,v}_{i+1}]$$

$$[T^{u,v}_i] = [t^{u,v}_{21,i}] [S^{u,v}_{i+1}] + [t^{u,v}_{22,i}] [T^{u,v}_{i+1}]$$

(21)

$$\{Q^w_i\} = [t^w_i] \{Q^w_{i+1}\}$$

Thus, after computing Eq. 21 for i = N-1 through 1, the compliance at the head can be obtained as

$$[F^{u,v}] = [S^{u,v}_1] [T^{u,v}_1]^{-1}$$

(22)

$$F^w = \frac{Q^w_1(1)}{Q^w_1(2)}$$

where $\{Q^w_1\} = (Q^w_1(1) \; Q^w_1(2))^T$.

Computation Procedures

In the present approach, a compliance matrix of a pile group is constructed from F_* and F_o, which are in turn computed from the analyses of single piles attached respectively to Winkler springs k_* and k_o. The computations of F_* and F_o are carried out in the following manner:

1. Compute $[k_b^{u,v}]$ and k_b^w.

2. Compute $[T_N^{u,v}]$, $[S_N^{u,v}]$ and $\{Q_N^w\}$ from Eq. 19.

3. Compute $[T_i^{u,v}]$, $[S_i^{u,v}]$ and $\{Q_i^w\}$ for i = N-1 through 1 using Eq. 21.

4. Compute $[F^{u,v}]$ and F^w from Eq. 22.

In the computations of $[T^{u,v}]$, $[S^{u,v}]$ and $\{Q^w\}$, the matrices $[t_{11}^{u,v}] \sim [t_{22}^{u,v}]$ and $[t^w]$ are obtained from the equations given in Appendix I, where k_o and k_* are given respectively by Eqs. 12 and 11. The matrix $[k_b^{u,v}]$ and k_b^w can be computed assuming that the pile-tip is supported by the soil column surrounded by a soil medium.

When the group has a large number of piles, the computations of F_* for all possible distances between pairs of piles in the group become tedious. It may be convenient to construct "amplitude of F_{ij} versus R" and "phase shift of F_{ij} versus R" curves first. When the soil medium is homogenous, the F_{ij} values can be directly obtained by substituting the amplitude ratios and phase shifts found in Figs. 5 and 6 into

$$F_{ij} = |F_o| \alpha (\cos \gamma - i \sin \gamma) \tag{23}$$

where α = amplitude ratio; $-\gamma$ = phase shift; and $|F_o|$ = amplitude of F_o. The parameter K_r in Figs. 5 and 6 are defined as

$$K_r = \frac{EI}{E_s L^4} \qquad \text{for flexural vibration}$$

$$\tag{24}$$

$$K_r = \frac{EA}{E_s L^2} \qquad \text{for axial vibration}$$

in which L = length of the pile; and E_s = Young's modulus of soil.

Illustrative Examples

A pile group considered here is made of 4x4 concrete piles attached to a common cap. The size and arrangement of the piles are as shown in Fig. 7. The soil profiles considered are also shown in Fig. 8.

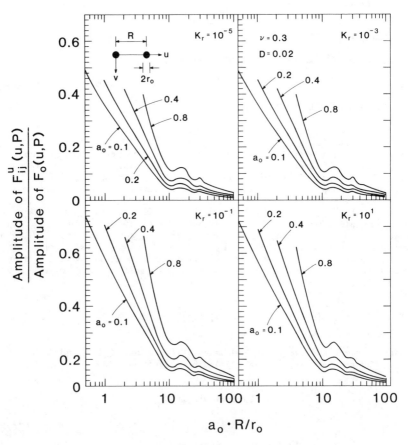

Fig. 5a Amplitude ratio for $F_{ij}^{u}(u,P)$

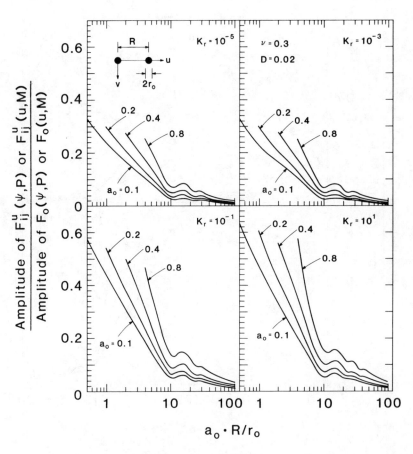

Fig. 5b Amplitude ratio for $F_{ij}^u(u,M)$ and $F_{ij}^u(\psi,P)$

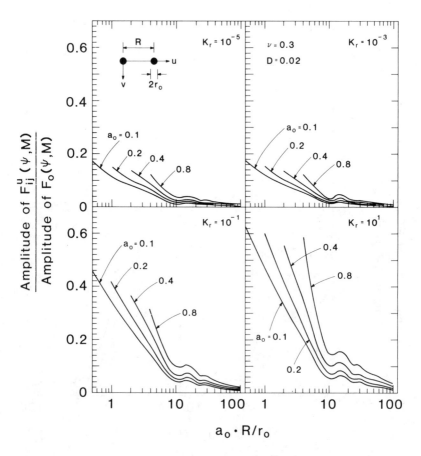

Fig. 5c Amplitude ratio for $F_{ij}^{u}(\psi,M)$

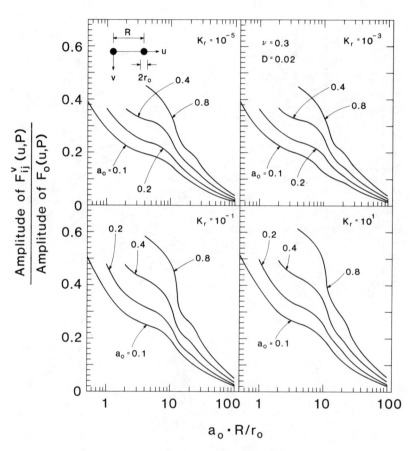

Fig. 5d Amplitude ratio for $F_{ij}^{V}(u,P)$

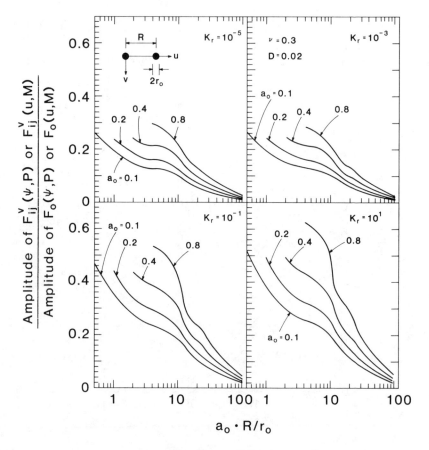

Fig. 5e Amplitude ratio for $F_{ij}^v(u,M)$ and $F_{ij}^v(\psi,P)$

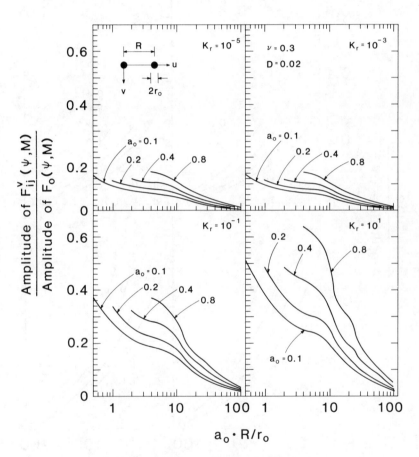

Fig. 5f Amplitude ratio for $F_{ij}^{v}(\Psi,M)$

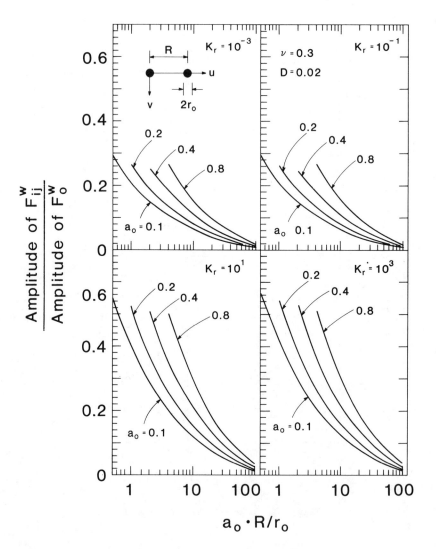

Fig. 5g Amplitude ratio for F_{ij}^w

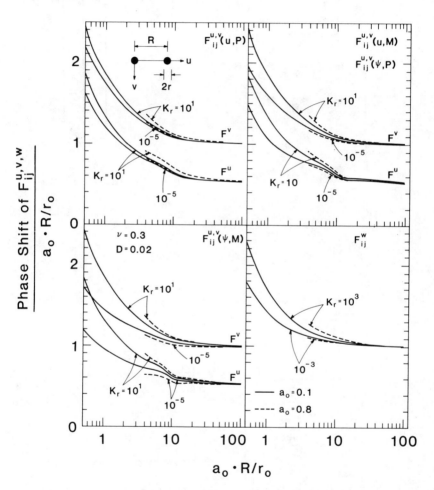

Fig. 6 Phase shift of $F_{ij}^{u,v,w}$

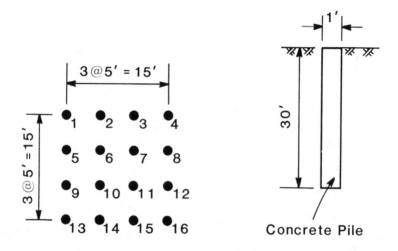

Fig. 7 Pile group considered in example

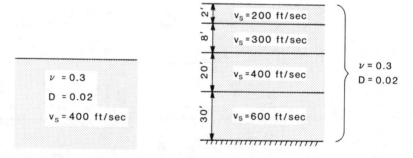

(a) Homogeneous Profile   (b) Nonhomogeneous Profile

Fig. 8 Soil profiles considered in example

Harmonic excitations with frequency 25 Hz are assumed to be applied at the center of the pile groups. The directions of the lateral and rocking excitations are in the x direction and around the y coordinate, respectively.

Homogeneous Soil Profile. - Assuming the Young's modulus of the concrete (E) as 467,975 ksf and determining the Young's modulus of the soil from the given Poisson's ratio and shear wave velocity of the soil medium, the K_r values are obtained as $K_r = 1.69 \times 10^{-3}$ for flexural vibration and $K_r = 6.32 \times 10^{-1}$ for axial vibration. The given frequency corresponds to $a_o = r_o \omega/v_s = 0.196$. With those K_r and a_o values, the amplitude ratios and phase shifts can be found from Figs. 5 and 6 for all possible pile pairs. For example, when the first and tenth piles are considered, the values found at $a_o \cdot R/r_o = 4.38$ are found as those given in Table 2. Substituting those values into Eq. 23, the $F_{ij}^{u,v,w}$ values are expressed as

$$
\begin{Bmatrix}
F_{ij}^u(u,P) \\[2mm]
F_{ij}^u(u,M) \\[2mm]
F_{ij}^u(\psi,P) \\[2mm]
F_{ij}^u(\psi,M)
\end{Bmatrix}
=
\begin{Bmatrix}
|F_o^{u,v}(u,P)| \; (-0.1702 + 0.0516\,i) \\[2mm]
|F_o^{u,v}(u,M)| \; (0.1178 - 0.0134i) \\[2mm]
F_{ij}^u(u,M) \\[2mm]
|F_o^{u,v}(\psi,M)| \; (0.0205 + 0.0754i)
\end{Bmatrix}
$$

$$
\begin{Bmatrix}
F_{ij}^v(u,P) \\[2mm]
F_{ij}^v(u,M) \\[2mm]
F_{ij}^v(\psi,P) \\[2mm]
F_{ij}^v(\psi,M)
\end{Bmatrix}
=
\begin{Bmatrix}
|F_o^{u,v}(u,P)| \; (0.1392 + 0.1932i) \\[2mm]
|F_o^{u,v}(u,M)| \; (-0.0676 - 0.1418i) \\[2mm]
F_{ij}^v(u,M) \\[2mm]
|F_o^{u,v}(\psi,M)| \; (0.0205 + 0.0754i)
\end{Bmatrix}
$$

$$
F_{ij}^w = |F_o^w| \; (0.0375 + 0.1314i)
$$

Eq. 7, with the above values and $\theta_{ij} = 63.43°$ leads to

$$
\left.
\begin{array}{l}
F^{xx}(1,10) \\[2mm]
F^{xx}(10,1)
\end{array}
\right\} = |F_o^{u,v}(u,P)|(0.0772 + 0.1649i)
$$

$$
\left.
\begin{array}{l}
F^{xx}(10,17) \\[2mm]
F^{xx}(1,26)
\end{array}
\right\} = |F_o^{u,v}(u,M)|(-0.0305 - 0.1161i)
$$

$$\left.\begin{array}{l} F^{xx}\ (17,10) \\ F^{xx}\ (26,1) \end{array}\right\} = |F_o^{u,v}(\psi,P)|(-0.0305 - 0.1161i)$$

$$\left.\begin{array}{l} F^{xx}(17,26) \\ F^{xx}(26,17) \end{array}\right\} = |F_o^{u,v}(\psi,M)\ |(0.0044 + 0.0595i)$$

$$\left.\begin{array}{l} F^{z}(1,10) \\ F^{z}(10,1) \end{array}\right\} = |F_o^{w}\ |(0.0375 + 0.1314i)$$

Since the conditions considered for the pile group and loadings do not produce translational motions in the y-direction nor rotational motions along the x coordinate, $[F^{xy}]$ $[F^{yx}]$ and $[F^{yy}]$ need not be computed.

Nonhomogeneous Profile. - The soil medium is divided into four horizontal layers as shown in Fig. 8. Analyses are conducted for single piles attached to the springs k_o and k_* to obtain F_o and F_*, respectively. The computation process to calculate $F_*^{u,w}$ for the first and tenth pile are described in the following, in

which the units are all in pounds and feet.

Layer 4

In order to obtain the soil stiffness at the pile-tip, the soil below the pile tips is assumed to be the system of vertical soil columns attached to springs as shown in Fig. 2. Replacing the two columns with a single column attached to a Winkler spring k_*, the pile-tip soil stiffnesses (the stiffnesses of the soil column at the head) for the computation of F_* are:

$$[k_b^u] = \begin{bmatrix} 2546 + 269i & 844 + 65i \\ 844 + 65i & 560 + 27i \end{bmatrix}$$

$$k_b^w = 2831 + 204i$$

Layer 3

The frequency parameter a_o is equal to 0.196. The soil stiffnesses k_* are computed from Eqs. 11, 13 and 14 as

$$\left\{\begin{array}{c} k_*^u \\ k_*^w \end{array}\right\} = \left\{\begin{array}{c} 1136 + 1115i \\ 841 + 9f12i \end{array}\right\}$$

With those k_*, the matrices [t] are computed from the equations in Appendix I as

$$[t_{11}^u] = \begin{bmatrix} 231 - 255i & -1735 - 63i \\ 330 + 246i & 230 - 255i \end{bmatrix}$$

$$[t_{12}^u] = \begin{bmatrix} -0.1765 - 0.0964i & -0.1646 - 0.0397i \\ 0.1646 + 0.0397i & 0.0755 + 0.0027i \end{bmatrix}$$

$$[t_{21}^u] = \begin{bmatrix} 3498892 + 625304i & -7405322 - 2934424i \\ 7405322 + 2934424i & -7584140 - 5650917i \end{bmatrix}$$

$$[t_{22}^u] = \begin{bmatrix} 230-255i & -330 - 246i \\ 1735 + 63i & 231 - 255i \end{bmatrix}$$

$$[t^w] = \begin{bmatrix} 16793 + 0.1766i & 0.6628 \times 10^{-4} + 0.2968 \times 10^{-5}i \\ 2733.4 + 77262i & 1.6793 + 0.1766i \end{bmatrix}$$

Substituting [t] and $[k_b]$ into Eq. 19, $[S_3^u]$, $[T_3^u]$ and $\{Q_3^w\}$ are obtained as

$$[S_3^u] = \begin{bmatrix} -430 - 621i & -1992.5 - 188i \\ 884 + 402i & 428 - 207i \end{bmatrix}$$

$$[T_3^u] = \begin{bmatrix} 3937439 - 338478i & -7366797 - 3315198i \\ 12796608 + 3221582i & -5818043 - 5658223i \end{bmatrix}$$

$$\{Q_3^w\} = \left\{\begin{array}{c} 4718 + 843i \\ 75805548 + 27456768i \end{array}\right\}$$

Layer 2

The frequency parameter a_o is equal to 0.262. The stiffness k_* and matrices [t] are computed similarly as those computed for Layer 3. Substituting [t], $[S_3^u]$, $[T_3^u]$ and $\{Q_3^w\}$ into Eq. 21, $[S_2^u]$, $[T_2^u]$ are computed as:

$$[S_2^u] = \begin{bmatrix} -15626 + 9966i & 41046 + 15170i \\ -5515 - 8784i & -14184 - 257i \end{bmatrix}$$

$$[T_2^u] = \begin{bmatrix} -80705546 - 22738174i & 57258445 + 6410466i \\ -161415204 - 96926675i & -1723684 + 873195559i \end{bmatrix}$$

$$\{Q_2^w\} = \begin{Bmatrix} 6627 + 1601i \\ 100920616 + 43399691i \end{Bmatrix}$$

Layer 1

The frequency parameter a_o is equal to 0.393. The matrices $[S_1^u]$ and $[T_1^u]$ and the vector $\{Q_1^w\}$ are computed from Eq. 21 as

$$[S_1^u] = \begin{bmatrix} 14382 + 37174i & 67054 + 4221i \\ -27059 - 18922i & -9585 + 13524i \end{bmatrix}$$

$$[T_1^u] = \begin{bmatrix} -9285640 - 4411205i & 102626354 + 95884538i \\ -337737579 - 130061629i & 137966858 + 243594028i \end{bmatrix}$$

$$\{Q_1^w\} = \begin{Bmatrix} 7188 + 1854i \\ 105406652 + 494137776i \end{Bmatrix}$$

Thus, F_* are obtained from Eq. 22 as

$$[F_*^u] = 10^{-4} \times \begin{bmatrix} 9.419 - 2.301i & -3850 + 0.506i \\ 2.850 + 0.507i & 1.540 - 0.135i \end{bmatrix}$$

$$F_*^w = 10^{-4} \times (0.627 - 0.118i)$$

Conclusions and Remarks

A simple approach to evaluate the compliance matrix is presented for pile groups subjected to vibrations at the heads. The approach requires numerical analyses only for single piles and thus is very efficient compared with convensional approaches which usually requires numerical analyses for the entire piles in a group simultaneously. Furthermore, charts are developed for quick evaluation of compliance matrices of pile groups for both flexural and axial vibrations. The frequency range considered include high frequencies such as those often encountered in machine foundations.

The major assumptions involved in the present approach occur in the idealization of the soil as a Winkler model and in the construction of the pile group compriance matrix from the analysis of all the two pile sets. The first assumption has been verified to be reasonable for frequencies higher than the fundamental frequency of the soil deposit (3,4). The second assumption has also been found to be reasonable (1,9,10).

Acknowledgment

The writers are grateful to Jun Otani, a graduate research assistant, for computing and plotting the results.

Appendix I- Matrices [t]

$$[t^w] = \begin{bmatrix} C & S \\ -S & C \end{bmatrix}$$

$$[t_{11}^{u,v}] = \frac{1}{2} \begin{bmatrix} CH+C & SH+S \\ SH-S & CH+C \end{bmatrix}$$

$$[t_{12}^{u,v}] = \frac{1}{2} \begin{bmatrix} -(SH-S) & CH-C \\ CH-C & SH+S \end{bmatrix}$$

$$[t_{21}^{u,v}] = \frac{1}{2} \begin{bmatrix} -(SH+S) & -(CH-C) \\ CH-C & SH-S \end{bmatrix}$$

$$[t_{22}^{u,v}] = \frac{1}{2} \begin{bmatrix} CH+C & -(SH-S) \\ -(SH+S) & CH+C \end{bmatrix}$$

where $S = \sin \lambda$; $C = \cos\lambda$; $SH = \sinh\lambda$; $CH = \cosh \lambda$; and

$$\lambda = \sqrt{(m\omega^2 - k^w)/(EA)} \quad \text{for } [t^w]$$

$$\lambda = \sqrt[4]{(m\omega^2 - k^{u,v})/(EI)} \quad \text{for } [t^{u,v}].$$

Appendix II - References

1. Kaynia, A. M. and Kausel, E., "Dynamic Behavior of Pile Groups," Proceedings, Second International Conference on Numerical Methods in Offshore Piling, I.C.E., 1982.

2. Nogami, T., Dynamic Stiffness and Damping of Pile Groups in Inhomogeneous Soil," ASCE Special Technical Publication on Dynamic Response of Pile Foundations: Analytical Aspect, 1980, pp. 31-52.

3. Nogami, T., "Dynamic Group Effect in Axial Responses of Grouped Piles," Journal Geotechnical Engineering Division, ASCE, No. GT2, Vol. 109, 1983, pp. 228-243.

4. Nogami, T., and Chen, H. S., "Behavior of Pile Group Foundations Subjected to Dynamic Loading," Proceedings, Fourth Canadian Conference on Earthquake Engineering, Vancouver, British Columbia, 1983, pp. 413-423.

5.   Nogami, T., and Chen, H. L., "Simplified Approach for Axial Pile
      Group Response Analysis," Journal of Geotechnical Engineering
      Division, ASCE, Vol. 110, No. G79, 1984, pp 1239-1255.

6.   Nogami, T., "Flexural Response of Grouped Piles Under Dynamic
      Loading," International Journal of Earthquake Engineering and
      Structural Dynamics, Vol. 13, No. 3, 1985.

7.   Poulos, H. G., "Analysis of Settlement of Pile Groups,"
      Geotechnique, Vol. 18, No. 4, 1968; pp. 449-471.

8.   Poulos, H. G., "The Behavior of Laterally Loaded Piles," Journal
      of Soil Mechanics and Foundations Division, ASCE, Vol. 97, No.
      SM5, 1971, pp. 733-751.

9.   Roesset, J. M., "Dynamic Stiffness of Pile Groups," ASCE Special
      Technical Publication on Analysis and design of Pile
      Foundations, 1984, pp. 263-286.

10.   Sanchez-Salinero, I., "Dynamic Stiffness of Pile Groups:
      Approximate Solution," Geotechnical Engineering Report GR 83-5,
      Dept. of Civil Engineering, University of Texas at Austin,
      1983.

11.   Sheta, M. and Novak, M., "Vertical Vibration of Pile Groups,"
      Journal of Geotechnical engineering Division, ASCE, Vol. 108,
      No. GT4, 1982, pp 570-590.

12.   Wolf, J. and Von Arx, G. A., "Impedance Function of a Group of
      Vertical Piles," Proceedings, ASCE Geotechnical Engineering
      Speciality Conference, Pasadena, 1978, pp. 1024-1041.

DYNAMIC STIFFNESS AND DAMPING OF FOUNDATIONS
BY SIMPLE METHODS

Ricardo Dobry[1], M.ASCE and George Gazetas[2], M.ASCE

ABSTRACT

The paper summarizes a number of simple models and methods developed by the authors to estimate the equivalent dynamic stiffness and damping of foundations. Four of these procedures are described in detail: a) a method for the equivalent vertical spring and dashpot of a rigid embedded foundation of arbitrary shape; b) a model for the radiation damping in all vibration modes of a rigid surface foundation of arbitrary shape, including useful asympotic upper bounds for high frequencies and long foundations; c) a model for the equivalent horizontal spring and dashpot of a pile in an arbitrarily layered soil deposit; and d) charts to obtain the fundamental frequency of an arbitrary soil profile on rock, which determines the range of low frequencies where no radiation damping exists in any of the foundation vibration modes.

It is concluded that simple methods, based on physically sound conceptual frameworks and calibrated with exact solutions, are feasible and should be encouraged. These simple methods are very helpful in performing parametric studies, checking the results of complicated computer codes and developing engineering insight into the expected dynamic behavior of foundation systems.

INTRODUCTION

A number of powerful computational techniques have been developed for calculating the dynamic stiffness and damping coefficients of foundations in linear soil. They include analytical approaches such as the integral transform method [17, 19, 36], several rigorous and approximate semi-analytical procedures such as the boundary element method [7], and dynamic finite elements [17, 24]. These techniques can be applied to the analysis of machine foundations and to other dynamic soil-structure interaction problems encountered in earthquake and offshore engineering. Although further extensions and refinements of these methods are continuously being published, they have already attained a great degree of sophistication and may today handle foundation and soil configurations of great complexity. Shallow and deep foundations can be solved for the different vibration modes, for exciting forces covering a wide range of frequencies, and for soil conditions ranging from a uniform deposit to an arbitrary layered profile on rigid rock. Recently, there has been a

---

[1]Professor of Civil Engineering, Rensselaer Polytechnic Institute, Troy, New York 12180.
[2]Associate Professor of Civil Engineering, Rensselaer Polytechnic Institute, Troy, New York 12180.

rapid development of methods for pile groups [18, 25, 31, 34, 40]. Summaries of these state-of-the art techniques and of results obtained with them have been presented by several authors [5, 9, 30].

However, the application of these methods and solutions to a specific engineering problem very often involves either using a specialized computer program, which may or may not be readily available, or searching through the technical literature for the most applicable curve or equation, and then adapting it as best as possible to the case at hand. This latter search can be a tedious task, as many research results have been published in a form not easily accessible to practicing engineers. Therefore, both alternatives may be be time consuming and expensive; this undoubtedly has limited the application of state-of-the-art procedures and results in projects which could have benefitted from them.

Even when the appropriate sophisticated code is used, the effort involved in getting one or two sets of usable results may be such that no time/budget is left at the end for necessary parametric studies. Such studies are of course critical for evaluating the effect of uncertainties in poorly known parameters (soil properties, quality of soil-foundation contact, etc.), and for exploring various design options.

An alternative has been the development of approximate simple methods, equations and charts, based on the results of more sophisticated formulations [29]. The classical example of such a simple method is the single-degree-of-freedom with frequency-independent parameters proposed in 1966 by Lysmer and Richart [23] for the vertical vibration of a rigid circular footing on an elastic halfspace. Similar simple systems or "analogs" were later developed for the other modes of vibration of the same circular footing, and these analogs are widely used today [28, 39]. Later, Kausel, Novak, Roesset, Veletsos and others have fitted equations and charts to their analytical results and have proposed simple procedures to take into account the variation with frequency of the equivalent dynamic springs and dashpots, and the influence on them of factors such as embedment and presence of a shallow rigid rock base [17, 26, 30, 36, 37].

In the last few years, the authors have contributed to this process by systematically developing a number of simple models and methods which can be used by practicing engineers without the need of a computer or with a minimum use of it. These models are based on sound, though simplified, analytical frameworks, with some of them taking advantage of basic principles of dynamics and wave propagation. In all cases, state-of-the-art solutions computed by the authors or retrieved from the literature have been used to verify these frameworks and to calibrate the final form of the proposed curves and equations. These methods developed by the authors can be used to define: (i) the dynamic stiffness (spring) and damping (dashpot) coefficients of surface foundations of arbitrary shape on deep uniform soil deposits, for all six modes of vibration [5, 6]; (ii) the dynamic vertical spring and dashpot of an embedded foundation of arbitrary base shape [14]; (iii) the dynamic vertical, horizontal and rocking damping coefficients for a surface foundation on a soil deposit with modulus increasing with an arbitrary power of depth [10, 13]; iv) the dynamic horizontal stiffness and damping of a pile embedded in a uniform deposit or in an arbitrarily layered soil profile, and including the effect of a rigid rock base under the soil [4, 12]; and (v) a method for

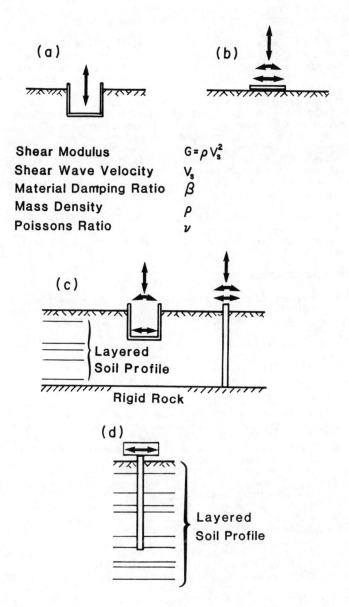

Fig. 1 - Soil-Foundation Systems Discussed in This Paper.

computing the fundamental frequency of an arbitrarily layered soil deposit on rock, and thereby determining the range of low excitation frequencies for which no radiation damping exists [3, 15].

The most obvious application of these simple methods and models is for preliminary calculations and in the conceptual design stage. However, and depending on the circumstances of the project, they can also be used for final calculations and design, alone or in combination with well planned sophisticated computer runs. The use of simple methods may involve some loss of precision, but this can be more than compensated by their inherent simplicity and accessibility, their ease of use in parametric calculations, and, especially, by their role in developing the practicing engineer's insight and feeling for the problem. Last but certainly not least, they provide the engineer with a "yardstick" to check and understand the results of complex computer programs, thus avoiding the well known "black box" syndrome. Also, as will be shown later herein, some of the simple models show remarkable predictive power and can be used to establish new results and useful bounds.

The paper illustrates the approach by discussing several of these simple models and methods proposed by the authors for shallow and deep foundations, with emphasis on their common features and underlying basic concepts.

BASIC FORMULATION

Fig. 1 sketches the soil-foundation systems discussed in this paper. Figs. 1(a) and (b) are embedded and surface rigid foundations on a deep soil, subjected to dynamic vertical and horizontal forces and to rocking and torsional moments; Fig. 1(c) refers to **any** foundation system on **any** soil underlain by rigid rock, including the case of an arbitrarily layered soil profile. Fig. 1(d) is a fixed-head pile embedded in an arbitrarily layered soil profile and subjected to a dynamic horizontal force. In all cases, the soil is characterized by the properties listed in Fig. 1(a). (The mass density, ρ = total unit weight/acceleration of gravity.) In some cases the soil parameters are constant, while in others they may vary with depth. The soil is assumed linear; that is, for the special case of $\beta = 0$, it behaves as a linearly elastic material.

For any foundation system and vibration mode, the soil can be replaced, for the dynamic analysis of the foundation and superstructure, by an equivalent dynamic spring $\overline{K}(\beta)$ and dashpot $C(\beta)$, selected to produce the same displacement of the rigid foundation (or pile cap). Fig. 2 sketches the vertical spring and dashpot, $\overline{K}_z(\beta)$ and $C_z(\beta)$, for a steady-state dynamic vertical force $V = V_0 \sin\omega t$ of frequency $\omega(\text{rad/s}) = 2\pi f$, where f is in cps. $\overline{K}_z(\beta)$ and $C_z(\beta)$ are selected to match the steady-state vertical displacement of the foundation, u_z, both in magnitude and phase. In the general case, there are six spring/dashpot pairs corresponding to the six modes of vibration of a rigid body: one vertical $[\overline{K}_z(\beta), C_z(\beta)]$; two horizontal $[\overline{K}_x(\beta); C_x(\beta); \overline{K}_y(\beta), C_y(\beta)]$; two rocking $[\overline{K}_{rx}(\beta), C_{rx}(\beta), \overline{K}_{ry}(\beta), \overline{C}_{ry}(\beta)]$; and one torsional $[K_t(\beta), C_t(\beta)]$. All these springs and dashpots are different for different foundation systems, for different soil properties and for the various vibration modes. Specifically, they do incorporate the effect of the material damping ratio, β, and they generally vary with the excitation frequency ω.

It must be noticed that, once the values of $\overline{K}(\beta)$ and $C(\beta)$ are determined for the case at hand, they can be used to represent the soil in the dynamic analysis of the structure, whether the structure supported by the foundation is rigid or flexible, whether it has only a few or many degrees of freedom. That is, the perfect rigidity assumed for the soil-foundation contact in Figs. 1(a) and (b), and for the pile cap in Fig. 1(d), should not be confused with any limiting assumption, not made here, about the degree of flexibility of the structure itself.

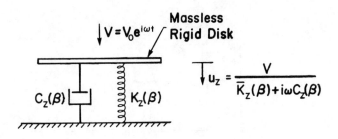

Fig. 2 - Equivalent Dynamic Spring and Dashpot for
Vertical Mode of Foundation Vibration.

There are two sources of energy dissipation in soil-foundation systems which contribute to the equivalent soil dashpot $C(\beta)$. One is the radiation or geometric damping associated with energy carried away from the foundation by stress waves traveling in the soil; this damping would exist even if the soil were perfectly elastic. The other is the hysteretic internal dissipation of energy within the soil, characterized by the value of the soil material damping ratio β. If β is a constant independent of depth (almost all cases discussed in this paper), it may be convenient to determine first a dynamic stiffness coefficient K and a radiation dashpot coefficient C by assuming that the soil is perfectly elastic and $\beta = 0$. Once this is done, the desired $\overline{K}(\beta)$ and $C(\beta)$ are readily obtained for the corresponding vibration mode by means of the general expressions:

$$\overline{K}(\beta) = \overline{K} - \omega\,\underline{C}\,\beta \qquad (1)$$
$$C(\beta) = C + 2\,\overline{K}\,\beta\,/\,\omega \qquad (2)$$

Eqs. 1 and 2 are based on the correspondence principle of viscoelasticity [22]. If β varies with depth, these expressions are not valid and a different procedure is required, as illustrated later herein when discussing the case of Fig. 1(d).

A MODEL FOR THE EQUIVALENT VERTICAL SPRING OF A RIGID EMBEDDED FOUNDATION OF ARBITRARY SHAPE

This is the case sketched in Fig. 1(a) and shown in more detail in Fig. 3. A rigid foundation of solid base area A_b and arbitrary shape, is embedded to depth D in a reasonably homogeneous and deep soil deposit. The possibility of partial sidewall contact is allowed for; ie., the total sidewall-soil contact area is $A_s \leq$ (D • perimeter of base area). The foundation is excited by a harmonic vertical force V_0 sinωt. In what follows, a simple procedure is described to obtain the dynamic vertical spring coefficent, $\bar{K}_z(\beta)$; the determination of the corresponding dashpot $C_z(\beta)$ is discussed in the next section.

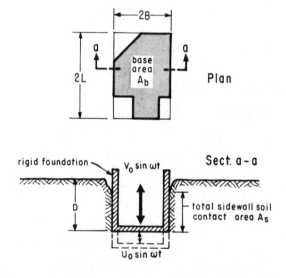

Fig. 3 - Rigid Embedded Foundation of Arbitrary Shape Under Vertical Vibration.

The first step is to determine the static spring coefficient, K_{emb}, corresponding to ω = 0 and β = 0. It can be shown [14] that:

$$K_{emb} = K_{sur} \cdot I_{tre} \cdot I_{wall} \tag{3}$$

where K_{sur} is the static spring of the same foundation placed at the surface of the soil; $I_{tre} \geq 1$ and $I_{wall} \geq 1$ are stiffening factors incorporating the "trench" and "wall" effects. The trench factor I_{tre} corresponds to the increase in stiffness when the foundation is placed at the bottom of a trench of depth D, but without any sidewall contact. The sidewall factor I_{wall} is directly related to the sidewall contact area

A_s. The following approximate expressions were obtained from results of rigorous numerical solutions: $K_{sur} = 2GLS_z/(1-\nu)$; $I_{tre} = 1 + D[1 + (4/3)(A_b/4L^2)/(21B)]$; $I_{wall} = 1 + 0.19(A_s/A_b)^{2/3}$. Therefore, the desired static spring K_{emb} is:

$$K_{emb} = \frac{2GLS_z}{1-\nu} \cdot \left[1 + \frac{1}{21}\frac{D}{B}\left(1 + \frac{4}{3}\frac{A_b}{4L^2}\right)\right] \cdot \left[1 + 0.19\left(\frac{A_s}{A_b}\right)^{2/3}\right] \quad (4)$$

in which S_z is the function of the base area shape factor $A_b/(4L^2)$ shown in Fig. 4, and the rest of the parameters of the equation are defined in Figs. 3 and 4.

The "data points" in Fig. 4 correspond to rigorous elasticity solutions compiled by the authors for a variety of base area shapes, ranging from square and circular to long rectangular foundations, and including odd shapes such as triangles, ellipses, rhomboids and hexagons [5].

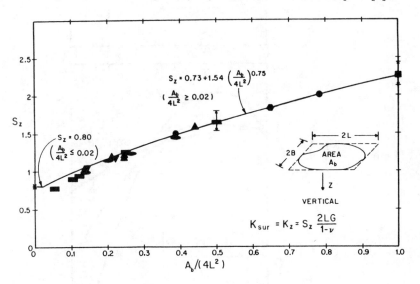

Fig. 4 - Vertical Surface Static Spring Parameter S_z Versus Base Shape [5].

If there is doubt about the quality of the sidewall contact, the engineer may want to use a smaller value of A_s in Eq. 4.

Once the static stiffness has been determined by Eq. 4, it is possible to obtain the dynamic spring $\overline{K}_{emb}$ at the frequency of interest and for assumed elastic soil ($\beta=0$). Fig. 5 plots the variation with dimensionless frequency $a_0 = \omega B/V_s$ of the normalized **surface** dynamic stiffness, $k_{sur} = \overline{K}_{sur}/K_{sur} = (\overline{K}_z/K_z)_{sur}$. Included in this figure are plots for the two extreme cases of unsaturated soil, $\nu \approx 0.3$ to 0.4, and

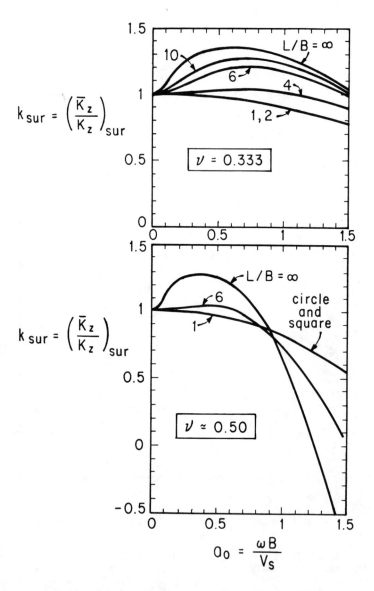

Fig. 5 - Dynamic Surface Vertical Spring Coefficient
Versus Frequency and Base Shape [5].

saturated clay, $\nu \approx 0.5$, and for a variety of foundation shapes defined by the aspect ratio of the circumscribed rectangle, L/B. Finally, the "elastic" ($\beta = 0$) dynamic spring coefficient of the embedded foundation, $\overline{K}_{emb}$ can be obtained as:

$$\overline{K}_{emb} = K_{emb} k_{emb} = K_{emb} \cdot (k_{emb}/k_{sur}) \cdot k_{sur} \tag{5}$$

where K_{emb} is the static value given by Eq. 4, k_{sur} can be read at the corresponding frequency in Fig. 5, and k_{emb}/k_{sur} is obtained with the help of the following expressions [14]:

Unsaturated soil ($\nu = 0.3$ to 0.4) and $a_o \leq 1.5$:

$$k_{emb}/k_{sur} = 1 - 0.09(D/B)^{3/4} \cdot a^2_o \tag{6}$$

Saturated clay ($\nu \simeq 0.5$) and $a_o \leq 1.5$:

$$k_{emb}/k_{sur} = \begin{cases} 1 - 0.08(D/B)^{3/4} \cdot a_o^2 & \text{if } L/B \simeq 1 \\ \\ 1 - 0.35(D/B)^{1/2} \cdot a_o^{3.5} & \text{if } L/B \geq 4 \end{cases} \tag{7}$$

The reader is cautioned that Eq. 7 for saturated clay is based on a smaller number of analytical results than those used to support Eqs. 4 and 6; thus, Eq. 7 should be used with some caution.

MODEL FOR THE RADIATION DAMPING OF A RIGID SHALLOW FOUNDATION OF ARBITRARY SHAPE

This section presents an analytical framework and a derived simple method for estimating the radiation damping coefficients, C, of a **surface** foundation of arbitrary shape, for **all** six modes of vibration (Fig. 1b). The method is then extended to include the calculation of the **vertical** radiation damping coefficient, $C_z = C_{emb}$, of an embedded foundation of arbitrary shape (Fig. 1a); this supplements the method for the stiffness coefficient K_{emb} presented in the previous section. The soil deposit is assumed to be elastic and reasonably deep and uniform. The contribution of soil material damping can be easily accomodated by means of Eq. 2.

Fig. 6 sketches the solid rigid area of arbitrary shape A_b, which constitutes the contact between foundation and soil surface. The figure defines the coordinate system and the six degrees of freedom associated with the radiation damping coefficients (dashpots):

C_x : Horizontal radiation dashpot parallel to the **long** axis of the foundation.

C_y : Horizontal radiation dashpot parallel to the **short** axis of the foundation.

C_z : Vertical radiation dashpot.

C_t : Torsional radiation dashpot.

C_{ry} : Rocking radiation dashpot for **motions** parallel to the **long** axis of the foundation.

C_{rx} : Rocking radiation dashpot for **motions** parallel to the **short** axis of the foundation

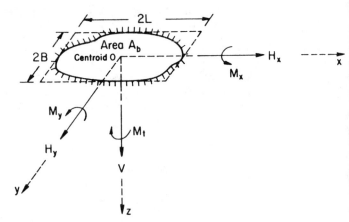

Fig. 6 - Rigid Surface Foundation of Arbitrary Shape: Vibration Modes.

First, a **high-frequency** model is developed for these six radiation dashpots, based on wave propagation concepts. It is well known [5, 23] that for a one-dimensional elastic wave (wave in a rod, plane body wave in an infinite space), $C=\rho VA$ is a perfect viscous dashpot "analog" for any wave shape and frequency. That is, the rod or space can be replaced by dashpot C, which fully absorbs the energy of the wave and thus simulates perfectly the radiation damping phenomenon. In the previous expression, ρ = mass density of elastic rod or space, V = propagation velocity of the wave, and A = cross-sectional area normal to the direction of propagation. This concept can be applied directly to the three translational (vertical and horizontal) modes, as in these modes the rigid foundation acts very much like the wave front of a plane wave, with all points within the area A_b moving in phase and having the same displacement at all times. When the footing is vibrating horizontally, mainly shear waves are generated in the soil and $V = V_s$. When the footing is vibrating vertically, compression-extension waves are generated into the soil, and in the immediate vicinity of the footing, $V \approx V_{La} = 3.4 V_s / [\pi(1-\nu)]$. These compression-extension waves **are not** compression body waves (P-waves) of the traditional type, and **do not** propagate with a velocity $V_p = V_s[2(1-\nu)/(1-2\nu)]^{1/2}$. The reason for this is that P-waves travel without lateral straining of the soil, while the compression-extension waves propagating under the footing do induce lateral strains. The authors have found that the "Lysmer's Analog Wave Velocity", $V_{La} < V_p$ is a good representation of this apparent propagation speed close to the foundation [5, 13].

All of this could suggest that the 1-D dashpot analog ρAV previously described could be used for the translational modes, with the appropriate wave velocity V_s or V_{La}. However, this is not yet the case: at low and intermediate frequencies the waves radiated by the foundation into the soil propagate in both vertical and nonvertical directions, and the phenomenon **is not** 1-D. However, as the frequency increases and the wavelength shortens, the phenomenon tends to become 1-D due to destructive wave interference, and indeed, $C \to \rho VA$. Therefore, at **high frequencies**, and for any solid foundation shape:

$$\left.\begin{array}{l} C_x \to \rho \, A_b \, V_s \\[6pt] C_y \to \rho \, A_b \, V_s \\[6pt] C_z \to \rho \, A_b \, V_{La} = \rho \, A_b \{3.4/[\pi(1-\nu)]\}V_s \end{array}\right\} \qquad (8)$$

An extension of the same basic concept allows obtaining high-frequency approximate values for the other three rotational (torsional and rocking) radiation dashpots. At these high frequencies and small wavelengths, the different points within the foundation contact area A_b act as independent sources radiating 1-D waves into the soil. These are 1-D **shear** waves when the footing is vibrating in **torsion**, and 1-D **compression-extension** waves for **rocking** vibration. This is equiv-alent to assuming that the soil can be replaced by a dynamic Winkler me-dium having distributed dashpots $\rho V_s dA_b$ (torsion) and $\rho V_{La} dA_b$ (rocking), with each of these distributed dashpots corresponding to an element of contact area dA_b. After integrating all these elementary dashpots throughout the contact area A_b, the desired total radiation dashpots are obtained at high frequencies and for the rotational modes:

$$\left.\begin{array}{l} C_t \to \rho \, V_s \, J \\[6pt] C_{ry} \to \rho V_{La} \, I_y \\[6pt] C_{rx} \to \rho \, V_{La} \, I_x \end{array}\right\} \qquad (9)$$

where J is the polar moment of inertia of the contact area A_b around the vertical axis, and I_y, I_x are the area moments of inertia around the horizontal axes y, x. Again, these high frequency values of C_t, C_{ry} and C_{rx} given by Eqs. 9 should be valid for any foundation shape.

The validity of this high-frequency model for the six radiation dashpots, presented in Eqs. 8-9, was confirmed by the authors in Ref. 5. For that purpose, detailed comparisons were performed with exact analyt-ical and numerical results obtained by Kausel [17], Luco and Wong [19, 20, 41], Dominguez and Roesset [7], Rücker [33], Veletsos, [36, 38], the authors and others. Figs. 7-10 present the curves obtained from these comparisons, in the form of plots of C/ρVA (translational modes), $C_t/\rho V_s J$ (torsion) and $C/\rho V_{La}I$ (rocking) for different shaped foundations [5].

Indeed, **all** curves in Figs. 7-10 tend to have values close to unity at high frequencies, thus confirming the high frequency model of Eqs. 8-9. The case not included in the figures is the horizontal dashpot in the long direction of the foundation, C_x, for which all available information in-dicates that the high frequency approximation, $C_x \approx \rho V_s A_b$, is valid at both low and high frequencies [5, 7].

The general shapes of the curves in Figs. 7-10, as well as the in-fluence of L/B on them, are as could be expected from wave propagation and wave interference considerations, The plots for C_t and C_{ry} in Figs. 9 and 10(b) are of special interest, because an extension of the high frequency model allows establishing upper bounds for these rotational dashpots. In both figures, when L/B is larger, the curves plot higher and they approach unity at lower frequencies. This is reasonable for these torsional and rocking (with motions along the length of the footing) vibrations. For long foundations having a large L/B, both torsional and rocking dissipations of energy are controlled by waves originating at the ends of the foundation, located at a distance of about L from the center. As L/B increases, the value of a_0 for which the two ends start behaving as independent wave sources, and at which the high frequency (short wavelength) approximation is valid, should decrease. By extending the

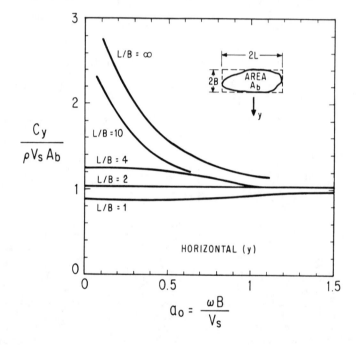

Fig. 7 - Horizontal Radiation Dashpot C_y Versus Frequency for Surface Foundations of Different Base Shapes [5].

same argument, the larger the value of L/B, the closer to zero should be the frequency at which the small wavelength approximations, $C_t/\rho V_s J = 1$ and $C_{ry}/\rho V_{La} I_y = 1$ are valid. Therefore, in the limit, for L/B = ∞, a jump from 0 to 1 should occur at $a_0=0$. This jump has been indicated in Figs. 9 and 10(b); these lines, in addition to providing values for the strip footing, constitute useful upper bounds for the values of C_t and C_{ry} of long foundations. The development of these upper bounds, based on the high-frequency model and the simple analytical framework just discussed, illustrates the fact that simple methods, by providing additional insight, can be used to produce new and useful results, not available or difficult to generate with more exact techniques.

Once the value of the vertical radiation dashpot coefficient $C_z = C_{z,sur}$ has been established for the surface foundation of area A_b and arbitrary shape, it is not difficult to determine the coefficient $C_{z,emb}$ for the corresponding embedded foundation. Due to the increased contact with the soil provided by the embedment, $C_{z,emb} > C_{z,sur}$; furthermore, the additional energy is dissipated mainly in the form of horizontally propagating shear waves transmitted to the soil by the sidewall contact

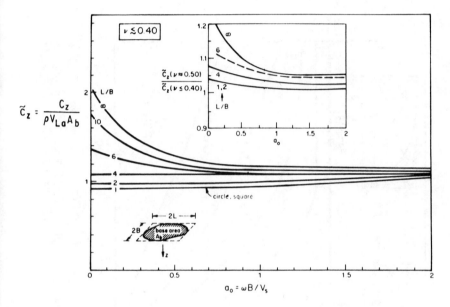

Fig. 8 - Vertical Radiation Dashpot C_z Versus Frequency for Surface Foundations of Different Base Shapes [5].

area A_s. Therefore, at high frequencies the surface dashpot $C_{z,sur}$ should be increased by simply adding a "sidewall contact" dashpot $\rho V_s A_s$. Analytical results [14] show that this is in fact a good approximation at both low and high frequencies. Thus:

$$C_{z,emb} = C_{z,sur} + \rho V_s A_s \qquad (10)$$

Eq. 10, which has the same general form but is different in detail from an expression originally proposed by Novak and Beredugo [26], is in good agreement with available solution for circular, square and rectangular foundations for D/B up to 2 [2,7]. Also, recent work by Chen, Roesset and Tassoulas [1] on a model of embedded walls having partial contact with the soil fully confirms the fact that the second term in Eq. 10 is proportional to the contact area A_s. Again, any doubt the engineer have about the quality of the sidewall contact with the soil can be incorporated by simply reducing the area A_s in Eq. 10.

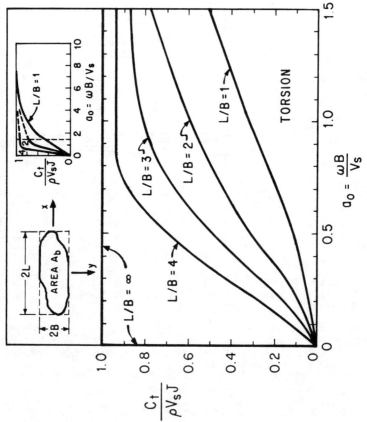

Fig. 9 – Torsional Radiation Dashpot C_t Versus Frequency for Surface Foundations of Different Base Shapes [5].

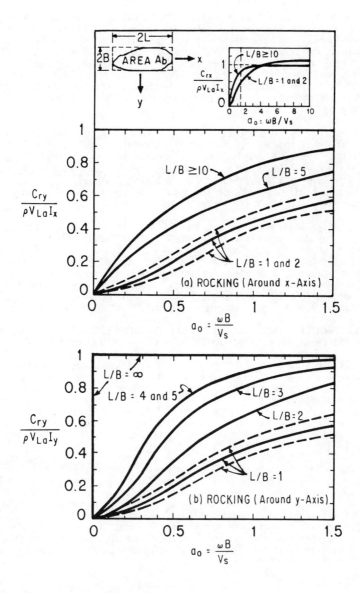

Fig. 10 - Rocking Radiation Dashpots C_{rx} and C_{ry} Versus Frequency for Surface Foundations of Different Base Shapes [5].

The authors have also developed another simple model for the horizontal, vertical and rocking radiation dashpots of circular and strip surface foundations on homogeneous and inhomogeneous deep soil [10, 13; see also 36 for previous work by Veletsos]. Based on this model, verified by some dynamic finite element calculations and by analytical results recently published by Wong and Luco [42], it is possible to state some general conclusions about the radiation dashpots of a surface foundation of arbitrary shape on inhomogeneous soil, the stiffness of which **increases** with depth:

1.   At high frequencies (small wavelengths), the foundation only "sees" the soil immediately beneath it, and the high-frequency Eqs. 8-9 are valid if the wave velocities at the surface of the soil, $V_{s,o}$ and $V_{La,o}$, are used. That is, at high frequencies, C_x and $C_y \rightarrow \rho A_b V_{s,o}$; $C_z \rightarrow \rho A_b V_{La,o}$; $C_t \rightarrow \rho J V_{s,o}$; $C_{ry} \rightarrow \rho I_y V_{La,o}$; and $C_{rx} \rightarrow \rho I_x V_{La,o}$, quite independently of the details of the foundation shape and of the exact law of increase of V_s and V_{La} with depth.

2.   At all frequencies and for all vibration modes, a homogenous halfspace of wave velocities $V_{s,o}$ and $V_{La,o}$ has larger values of the radiation damping coefficients than the corresponding inhomogeneous halfspace. This is intuitively reasonable, as the increase of stiffness with depth acts as a barrier, reflecting waves back to the foundation and thus allowing less energy to be propagated away from the footing than in the homogenous case. Therefore, the homogeneous values provided by Figs. 9-12 can be used as upper bounds for cases of inhomogeneous profiles.

SIMPLE MODEL FOR THE EQUIVALENT HORIZONTAL SPRING AND DASHPOT OF A PILE IN AN ARBITRARILY LAYERED SOIL DEPOSIT

This section presents a simple and practical method to estimate the equivalent horizontal spring and dashpot at the top of a pile embedded in an arbitrarily layered soil deposit, which may or may not be underlain by a rigid rock base (Fig. 1d). The soil layers may possess different stiffnesses, Poisson ratios and material damping ratios. The method is based on the analytical framework developed in Refs. 4, 5 and 13, and has been verified by comparison with more exact dynamic finite element analyses [12].

The pile can be either floating or end-bearing, and it is assumed to be a fixed-head "flexible" pile embedded in the layered soil. That is, due to the restriction imposed by the pile cap, there is no rotation at the top, and the pile is long enough or flexible enough so that only part of its length contributes to the dynamic response at the top. This last constraint is not very restrictive in practice, as most piles are longer than the 10-15 diameters required to comply with it. The pile may have a variety of cross sections, including circular, pipe, square, rectangular and concrete-filled steel pipe.

The procedure involves three basic steps, as follows:

1.   The horizontal displacement profile, $y_s(z)$ and the static value of the spring coefficient, $K_h = P_0/y_s(0)$, are obtained by subjecting the pile to the static horizontal force P_0 at the pile head. Any reasonable procedure   can   be   used   for   these   estimates,   including   a

beam-on-Winkler-foundation formulation with distributed linear or non-linear (p-y curves) springs along the length of the pile; a static finite element code; a full-scale lateral pile load test of an instrumented pile in the field; etc. In most cases, the dynamic coefficients $\bar{K}_h$ and $\bar{K}_h(\beta)$ are not very different from the static value K , and thus can be readily estimated once K_h is known. Fig. 11, reproduced from Ref. 12, contains plots of $\bar{K}_h(\beta)/K_h$ versus a_0, for a wide range of soil profiles and pile stiffnesses, and for a typical value of $\beta=0.05$. These plots are used in the selection of $\bar{K}_h(\beta)$ at different excitation frequencies. Note in Fig. 11 that for soil profiles on rigid rock the value of $\bar{K}_h(\beta)$ has always a dip at a_{os}, corresponding to the fundamental frequency in shear of the profile, f_s; this dip becomes more pronounced as β decreases. A simple method for obtaining the values of f_s and a_{os} is discussed in the following section.

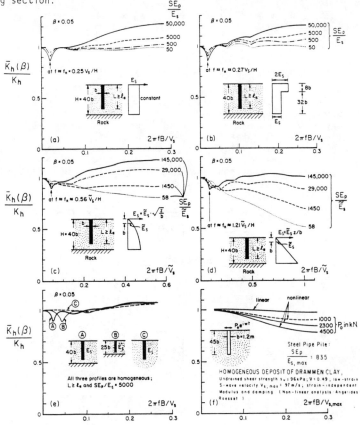

Fig. 11 - Equivalent Horizontal Dynamic Spring Coefficient of Pile, $\bar{K}_h(\beta)$, versus Frequency, from 3-D Finite Element Analyses. (E_s, E_p are Young's Moduli of Soil and Pile; S=Pile Cross-Section Shape Factor = 1 for Circle) [12].

2. At each elevation, two parallel soil dashpots are assumed attached to the pile. The parameters of these distributed dashpots are c_m and c_r per unit length of pile and represent, respectively, the material and radiation damping effects. Both c_m and c_r depend on soil properties and are different for the different layers. $c_m \approx 2k\beta/\omega$, where k is related to the stiffness of the soil layer, and β can be either specified a priori, or can be related to the shear strain induced in the soil by the static deflection of the pile computed in step 1 [12]. The value of the distributed radiation damping coefficent, c_r was obtained by the authors from the plane-strain 2-D model for wave propagation due to pile in-plane vibration, sketched in Fig. 12. The corresponding expressions obtained by the authors for c_r give results in good agreement with more exact calculations by Roesset and Angelides [32] and Novak et al. [27]:

$$c_r/(4B\rho_s V_s) = 1.67 \, a_o^{-1/4} \tag{11}$$

for a depth z < 5B; and

$$c_r/(4B\rho_s V_s) = 0.83 \left\{ 1 + [3.4/[\pi(1-\nu)]^{5/4} \right\} \cdot a_o^{-1/4} \tag{12}$$

for a depth z > 5B, where B is the radius or equivalent radius of the pile. A further correction of the values of c_r obtained with Eqs. 11-12 is necessary at low frequencies below or around the fundamental frequency in shear of the soil profile, $a_{os} = 2\pi f_s B/V_s$, if the soil is underlain by a rigid rock base. This is due to the fact that no radiation damping is possible for frequencies below f_s. It is suggested that $c_r = 0$ for $a_o < a_{os}$; c_r is computed from Eqs. 11-12 for $a_o \geq 2a_{os}$; and a straight-line transition is used between a_{os} and $2a_{os}$.

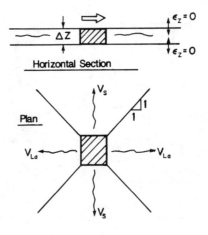

Fig. 12 - Plane Strain 2-D Radiation Damping Model
for Pile Under Horizontal Loading.

3. The overall dashpot coefficient $C_h(\beta)$ at the head of the pile, including both material and radiation damping effects, is obtained from the values of c_r and c_m distributed along the pile for the different layers (step 2), in conjunction with the static pile deflection profile $y_s(z)$ from step 1. The following simple energy-conservation expression is used:

$$C_h(\beta) = \int_0^L (c_r + c_m) \, Y^2_s \, (z) \, dz \tag{13}$$

in which L is the length of the pile, and $Y_S = y_S \, (z)/y_S(0)$ is the normalized static deflection profile.

The method was compared in Ref. 12 with results of dynamic finite analyses for several piles and soil profiles, with excellent agreement in all cases. One of these comparisons is reproduced in Fig. 13, for a stiff, large diameter concrete filled steel pile floating in a soil profile representative of the San Francisco Bay Area. Fig. 13(a) shows the soil profile, soil properties and calculated static deflection profile, while Fig. 13(b) presents the comparisons between $\overline{K}_h(\beta)$ and $C_h(\beta)$ computed using the simple method and those obtained from a 3-D dynamic finite element code.

FUNDAMENTAL FREQUENCY OF AN ARBITRARY SOIL PROFILE ON ROCK

The energy corresponding to radiation damping is generally carried away by body waves (which propagate in any direction and can exist far from any boundary) and surface waves (which propagate horizontally along the ground surface). When a rigid rock base exists at a certain depth below the soil, as sketched in Fig. 14, it prevents the propagation of the body waves (S and P) by reflecting them back to the foundation and to the ground surface; eventually their energy is either lost within the soil due to the material damping, or, at some distance from the foundation, gets converted into surface wave energy. Therefore, in the presence of a perfectly rigid rock base, only surface waves contribute to the radiation damping coefficients C in all vibration modes. It can be shown that no surface waves can exist in the presence of a rigid rock base at low frequencies f below the fundamental frequency of the soil in shear, f_s, which thus acts as a cutoff frequency [21]. Therefore, C=0 for all $f < f_s$. This conclusion is rigorously true for a perfectly elastic soil (β=0), and is approximately correct for the more practical case in which there is material damping. Also, it is a very general conclusion, valid for any foundation type (shallow foundation of any shape, embedded foundation, piers, floating and end-bearing piles, pile groups, etc.), and for all vibration modes. In practice, the fact that there is no radiation damping at low frequencies is especially important for the horizontal and vertical modes, for which the values of C in the absence of rigid rock can be very important. On the other hand, damping in torsion and rocking is usually quite small at low frequencies in many cases (see Figs. 9-10), and thus the presence of the rock may not make much difference. The fact that, indeed, C≈0 for $f < f_s$ has been verified for a number of cases by several authors [4, 11, 12, 17, 30, 35, 42].

Therefore, in the presence of rigid rock, it should be assumed C=0 for all modes at $f < f_s$, ($\omega_s < \omega_{os}, a_0 < a_{os}$). At those low frequencies the only source of damping available to the system is the material soil

(a)

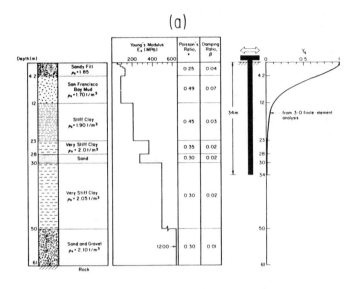

(b)

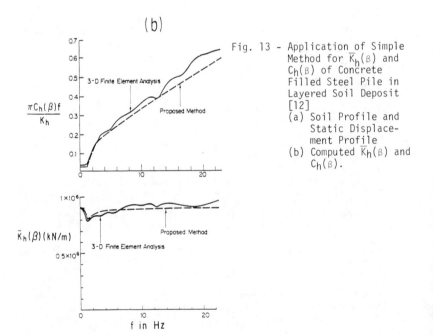

Fig. 13 - Application of Simple
Method for $\overline{K}_h(\beta)$ and
$C_h(\beta)$ of Concrete
Filled Steel Pile in
Layered Soil Deposit
[12]
(a) Soil Profile and
Static Displace-
ment Profile
(b) Computed $\overline{K}_h(\beta)$ and
$C_h(\beta)$.

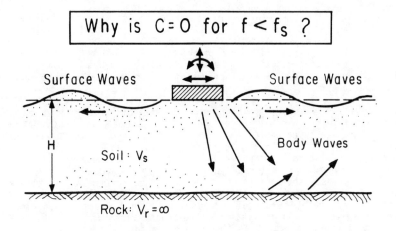

1. In elastic half–space ($H=\infty$) and far field, C is associated with 2 wave types:

$$\left.\begin{array}{l}\text{S–waves} \\ \text{P–waves}\end{array}\right\} \text{ body–waves}$$

$$\left.\begin{array}{l}\text{R–waves} \\ \text{L–waves}\end{array}\right\} \text{ surface–waves}$$

2. In layer over rigid rock, S and P waves do not exist in far field: only surface waves contribute to C.

3. For $f < f_s$, where $f_s = V_s/(4H)$, surface waves do not exist, and thus $C = 0$ in all vibration modes.

4. This is true generally for any soil profile over rigid rock:
   $C = 0$ in all modes at frequencies $f < f_s$, where
   f_s = fundamental frequency in shear of profile.

Fig. 14 - Radiation Damping Behavior at Low Frequencies, Foundation on Soil Overlying Rigid Rock.

damping ratio β, and the general Eqs. 1-2 become $K(\beta) = \overline{K}$ and $C(\beta) = 2\overline{K}\beta/\omega$.

It is then very convenient to have solutions and methods to estimate f_S of soil profiles likely to be encountered in practice. The two authors [3, 15] have published a number of these solutions, which are summarized in this section, including a simple method to estimate f_S of an arbitrarily layered soil profile on rock. Also, information is provided here on how to determine the radiation damping C_y for horizontal vibrations at low frequencies if the "base" of the deposit is not bedrock but simply consists of another, stiffer material. Finally, a general recommendation is also provided for a more precise estimate of the vertical and rocking damping behavior.

Once f_S is obtained for the soil profile of interest, it can be compared with the excitation frequency(ies) to verify if C obtained from the half-space solutions (i.e. from Figs. 7-10) must be modified or not. In this way, even if the foundation-soil system is complicated and the use of a sophisticated computer code is anticipated, useful information on a key dynamic parameter can be obtained at the outset. Furthermore, as in many cases the excitation frequency is determined by the machinery to be installed, and the soil profile, controlling f_S, is also fixed, the designer may be able to realize immediately that the radiation damping of the system is C=0 and that high resonances are possible; this realization is very important as the problem cannot be easily overcome by changes in the design of the foundation.

Rigorous solutions were obtained for the following cases of practical interest:

(a) Uniform layer on rigid rock: $f_S = V_S/(4H)$,

(b) Nonuniform layer of thickness H and stiffness increasing with a power of depth ($V_S = 0$ at the ground surface, $V_S = V_0 z^{p/2}$ at depth z, and p≤1.8). This case is representative of normally consolidated soil deposits, which typically have values of p between 0.5 and 1. The solution is expressed in the same standard form used for the uniform layer, $f_S = V_{eq}/(4H)$, where V_{eq} is selected at the representative depth z_{eq}. Some selected values of z_{eq}/H for various p obtained in Ref. 3 are given in Table 1.

(c) Nonuniform layer of thickness H and shear modulus G increasing or decreasing linearly with depth between G_0 (at z=0) and G_H (at z=H). G increasing linearly with depth is typical of many overconsolidated, rather stiff soil deposits. The solution is again expressed as $f_S = V_{eq}/(4H)$, where $V_{eq} = (G_{eq}/\rho)^{1/2}$, and G_{eq} is selected at the representative depth z_{eq}. Table 2 gives selected values of z_{eq} obtained in Ref. 3 for several values of the ratio $V_{SO}/(V_{SH}) = (G_0/G_H)^{1/2}$.

(d) Nonuniform layer of thickness H and shear wave velocity V_S increasing with depth in the form $V_S(z) = V_{SO}(1 + az)^m$, where m = 1/4, 1/2, 2/3 and 1, and a = positive constant characterizing the degree of inhomogenity. The solution is again expressed as $f_S = V_{eq}/(4H)$ where V_{eq} is the velocity at a depth z_{eq}. Fig. 15, reproduced from Ref. 15 shows the dependence of z_{eq} on the power m and the ratio $V_S(H/2)/V_{SO}$.

Table 2. Fundamental Frequency in Shear, f_s, of Nonuniform Layer with $G = G_o + (G_H - G_o) z/H$. (from Ref. 3)

$(G_o/G_H)^{1/2}$	z_{eq}/H
0.1	0.60
0.3	0.62
0.5	0.65
1	0.70
2	0.77
3	0.81
5	0.84
7	0.87

$G = \rho V_s^2$ = Shear Modulus
G_o = Modulus at Ground Surface
G_H = Modulus at Depth H
z = Depth
H = Layer Thickness

$$f_s = \frac{(G_{eq}/\rho)^{1/2}}{4H}$$
$$G_{eq} = G_o + (G_H - G_o) z_{eq}/H$$

Table 1. Fundamental Frequency in Shear, f_s, of Non-Uniform Layer with $V_s = V_o z^{p/2}$. (from Ref. 3)

p	z_{eq}/H
1/3	0.64
1/2	0.63
2/3	0.62
1	0.59
1.25	0.56
1.5	0.52
1.75	0.47

V_s = Shear Wave Velocity
z = Depth
H = Layer Thickness

$$f_s = \frac{V_{eq}}{4H}$$
$$V_{eq} = V_o z_{eq}^{p/2}$$

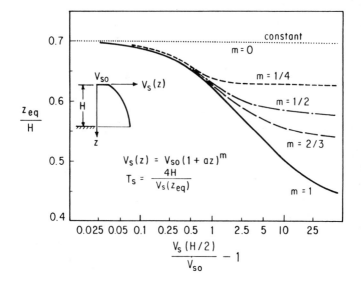

Fig. 15 - Fundamental Period T_s of Soil Deposit
on Rigid Rock [15].

(e) Two uniform soil layers. The chart developed in Ref. 3 for this
case is reproduced in Fig. 16. Unlike the solutions (a), (b), and (c)
above, which assume constant soil density with depth, in Fig. 16 the two
layers may have different mass densities ρ_A and ρ_B. The fundamental pe-
riods of the two individual layers, $T_A = 1/f_A = 4H_A/V_A$ and $T_B = 1/f_B = 4H_B/V_B$, must be computed first before entering the graph. The chart gives
the natural period of the two-layer system, T_s, from which $f_s = 1/T_s$ can
be readily computed.

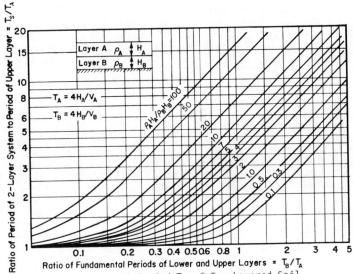

Fig. 16 - Fundamental Period T_S of Two-Layered Soil Profile on Rigid Rock [3].

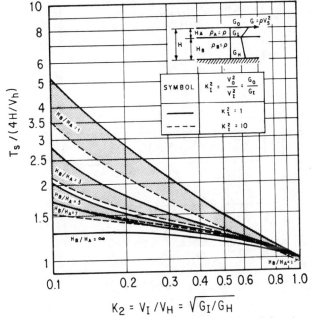

Fig. 17 - Fundamental Period T_S of Overconsolidated Crust Over Layer with Modulus Increasing with Depth [3].

(f) Two nonuniform soil layers. In the top layer, G decreases linearly with depth, while in the bottom layer G increases linearly with depth. The chart in Fig. 17, reproduced from Ref. 3, expresses the natural period $T_S = 1/f_S$ of the profile in terms of three parameters: two of them, K_1 and K_2, define the ratios of values of G at the boundaries of each layer, and the third is the ratio of thicknesses H_B/H_A.

In addition to these rigorous solutions for special cases of stiffness variation with depth, a simple approximate method can be used to estimate T_S and f_S for an arbitrarily layered soil profile on rigid rock, such as sketched in Fig. 18. This method is based on repeated application of the two-layer chart of Fig. 16, and it gives estimates of T_S and f_S within 10% of the exact values [3]. Tables 3 and 4 illustrate the use of this simple method with an example. The top two layers of the profile are first assumed to lie on rock and their combined period T_{1-2} is obtained using Fig. 16. In the example, $T_{1-2} = 0.068$ sec. Then, these top two layers are replaced by a single layer of thickness $H = H_A + H_B$ and $T = T_{1-2}$; this new top layer is combined with the third layer of the profile and the chart of Fig. 16 is used again to estimate the period of the top three layers of the original profile. In the example, as indicated in Table 4, $H_A = 8$ ft, $H_B = 5$ ft, $H_C = 7$ ft and $T_{1-2-3} = 0.09$ sec.

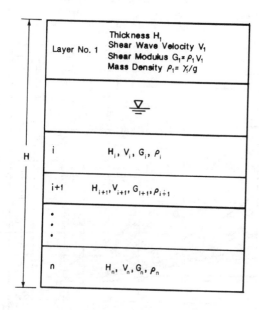

Fig. 18 - Layered Soil Profile.

The process is repeated until all soil layers have been considered and H coincides with the total thickness of the whole soil profile (H = 200 ft in the example). The last value of $T_{1-2-3-4...}$ approximates the desired T_S. In the example of Table 3, the estimated T_S = 0.59 sec, essentially identical to the exact value, T_S = 0.593 sec. The corresponding estimated fundamental frequency of the same profile is then f_S = 1/0.59 = 1.69 cps and ω = $2\pi/0.59$ = 10.6 rad/sec. Therefore, for any machine foundation placed on this profile and for exciting frequencies less than 1.7 cps, all modes of vibration of the foundation possess only material soil damping, and the radiation damping contribution should be taken as zero.

As mentioned before, all solutions presented above for f_S define a low frequency region $f < f_S$ where $C \approx 0$ for all vibration modes in the presence of rigid rock. However, the behavior for frequencies slightly above f_S is different in the cases of horizontal and torsion vibration, on one hand, and vertical and rocking vibration, on the other. Therefore, it is useful to discuss separately these two general situations, and also to examine the case in which the "base rock" is not perfectly rigid.

For **horizontal and torsional** modes and rigid rock, $C_x \approx C_y \approx C_t \approx 0$ for f $< f_S$, and C rapidly tends to approach the halfspace value as soon as the frequency becomes larger than f_S. This assumes that the base is a perfectly rigid rock with a shear wave velocity $V_r = \infty$. In practical situa-

Table 3. Layered Soil Profile for f_S Estimate.
          (from Ref. 3)

Layer	Thickness	V_S	Total Unit Weight γ	$\rho = \dfrac{\gamma}{g}$
1	8 ft	828 fps	105 pcf	3.26 lb-sec^2/ft^2
2	5	726	133	4.13
3	7	1039	120	3.73
4	8	825	120	3.73
5	5	951	137	4.25
6	65	1270	125	3.88
7	24	1065	127	3.94
8	16	1205	119	3.70
9	9	1071	138	4.29
10	7	1633	135	4.19
11	21	1223	138	4.29
12	25	2777	140	4.35
13(rigid rock)	∞	∞	---	----

ations in which the base material (V_r) is much stiffer than the overlying soil (V_s), and thus V_s/V_r is correspondingly very small, this is a reasonable assumption for which the solutions for f_s presented in this section are clearly applicable. This would be, for example, the case of a soft clay soil on sound granite rock. However, at many sites the "base" may be soft rock, or simply a soil layer significantly stiffer than those above it. The question then arises: how much stiffer must be the lower layer for the radiation damping of the foundation to be significantly affected in the low frequency range $f < f_s$? The authors developed the plot shown in Fig. 19 to help answer the previous question for the important case of horizontal vibration. The chart is based on analytical solutions presented in Refs. 11 and 42 for the case of a soil layer on "soft" rock sketched in Fig. 19(a), and for both strip and circular surface foundations. The cases analyzed correspond to a rock or hard layer at a relatively shallow depth (H/B = 2 and 3.55), and thus the figure should be used with some caution for cases in which the rock is deeper.

In Fig. 19(b), $C_y/C_{y,HS}$ is plotted versus f/f_s, where $C_{y,HS}$ is the horizontal radiation damping coefficient for a homogenous halfspace ($V_s/V_r = 1$), The plot shows a consistent behavior between circular (L/B=1) and strip (L/B=∞) foundations; this suggests that the chart may be used as a first approximation for the horizontal damping of arbitrary solid shapes having any aspect ratio L/B, including circular, square, rectangular and other shapes. The figure indicates that a base layer having a shear wave velocity twice that of the upper soil, $V_s/V_r = 0.5$, can decrease C_y to less than half of its halfspace value. Even if the wave velocity of the base layer is only about 30% higher, $V_s/V_r = 0.75$, a substantial reduction of C_y still takes place at low frequencies. Therefore, if a significantly stiffer layer is present in the soil profile of interest, it is recommended to estimate f_s for the soil above it using the techniques described in this section. Then, Fig. 19 can be used to estimate the reduction of C_y at low frequencies below f_s.

Table 4. Estimate of f_s for Soil Profile of Table 3 Using Simple Method Based on Fig. 16. Estimated $f_s = (0.59)^{-1} = 1.69$ Hz. (from Ref. 3).

Layers Considered From Ground Surface Down	H_A (ft)	H_B (ft)	H_A/H_B	V_A (fps)	V_B (fps)	T_A (sec)	T_B (sec)	T_B/T_A	T/T_A	T (sec)
1-2	8	5	1.6	828	726	0.039	0.028	0.72	1.75	0.068
1-2-3	13	7	1.9	726	1039	0.068	0.027	0.40	1.3	0.09
1 to 4	20	8	2.5	1039	825	0.09	0.039	0.43	1.45	0.131
1 to 5	28	5	5.6	825	951	0.131	0.021	0.16	1.15	0.151
1 to 6	33	65	0.51	951	1270	0.151	0.021	1.39	2.2	0.33
1 to 7	98	24	4.1	1270	1065	0.33	0.09	0.27	1.32	0.436
1 to 8	122	16	7.6	1065	1205	0.436	0.053	0.122	1.12	0.488
1 to 9	138	9	15.3	1205	1071	0.488	0.034	0.07	1.07	0.522
1 to 10	147	7	21.0	1071	1633	0.522	0.017	0.033	1.0	0.522
1 to 11	154	21	7.3	1633	1223	0.522	0.069	0.132	1.13	0.59
1 to 12	175	25	7.0	1223	2777	0.59	0.036	0.061	1.0	$0.59 = T_s$

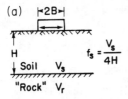

Symbol	Foundation Shape	H/B	V_s / V_r	Ref.
○	Circle	3.55	0.3–0.6–0.8	42
□	Strip	2	0.24–0.5	11

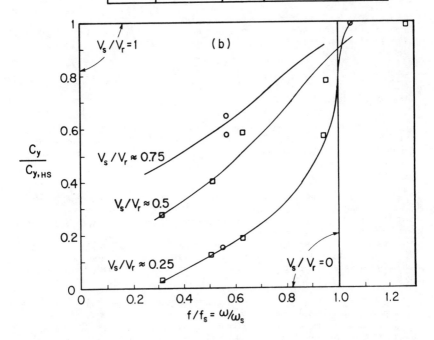

Fig. 19 - Horizontal Radiation Damping Coefficient, C_y, of Foundation on Soil Layer Underlain by "Flexible" Rock, as a Fraction of the Half-Space Value, $C_{y,HS}$.

For **vertical and rocking** modes and rigid rock, it is still true that $C_z \approx C_{rx} \approx C_{ry} \approx 0$ for $f < f_s$. However, unlike C_x, C_y and C_t, the vertical and rocking radiation damping coefficients do not tend to approach the halfspace value for $f > f_s$. Instead, they are still close to zero for a range of frequencies $f > f_s$, and C_z, C_{rx} and C_{ry} only start increasing when $f > f_c > f_s$, where $f_c > f_s$ is a different cutoff frequency which depends on f_s and on the Poisson's ratio of the soil, ν. From an examination of a number of analytical results of vertical and rocking response of both shallow foundations and piles on soil underlain by a stiff base [11, 30, 35, 42], the authors have concluded that f_c is approximately equal to the "fundamental frequency of the soil profile" which would be obtained if V_s were replaced by the "Lysmer's Analog Wave Velocity", $V_{La} = 3.4 V_s/[\pi(1-\nu)]$. It is interesting to note that f_c **is not** determined by the dilatational wave velocity of the soil, $V_p = [2(1-\nu)/(1-2\nu)]V_s$, but rather by the same compression-extension wave velocity, V_{La}, which was found previously to control the dissipation of radiated energy from vibrating footing and piles. If f_c were controlled by V_p, it would be $f_c = \infty$ at $\nu = 0.5$, which is not true. Therefore, all methods developed in this section for f_s can also be used to estimate f_c, by simply replacing V_s by V_{La} in the calculations. Once this is done, it can be assumed that $C_z = C_{rx} = C_{ry} = 0$ for $f < f_c$, and that C_z, C_{rx} and C_{ry} tend to increase toward the halfspace values for frequencies larger than f_c. For the very usual case in which the Poisson's ratio of the soil is constant with depth, f_c can be estimated from f_s of the soil profile by means of the simple expression:

$$f_c = 3.4/[\pi(1-\nu)] \, f_s \tag{14}$$

CONCLUSIONS

This paper makes the case for maximum use of simple models and methods to estimate the equivalent dynamic stiffness and damping of foundations. Based on the discussions and examples of simple methods presented through the paper, it is possible to draw the following conclusions:

1. The development of simple methods is feasible for both the stiffness and damping coefficients of shallow and deep foundations. These simple methods use as a starting point very simple but physically sound models and conceptual frameworks, and calibrate them by comparison with more exact analytical/numerical solutions. The use of basic wave propagation concepts has proven especially useful for the understanding and modeling of the radiation damping contribution to the different vibration modes.

2. The use of simple methods is convenient and should always be considered for the dynamic analysis and design of foundations, often in conjunction with more exact and sophisticated techniques. When compared with these techniques, simple methods are less precise but are simpler, more accessible, inherently suited for performing parametric calculations and for checking results of complicated computer codes, and very helpful in developing engineering insight into the problem.

3. At the very minimum, simple procedures are helpful as tools for interpolating between results of more exact solutions; this is shown very clearly by the methods for spring and dashpot coefficients of shallow foundations of arbitrary shape discussed in this paper. In addition, simple models can also exhibit remarkable predictive power, allowing the extrapolation of exact solutions to new situations, and thus establishing new results and useful bounds.

4. Therefore, the development of more simple methods and their systematic use by engineers should be encouraged.

REFERENCES

1. Chen, H. T., Roesset, J. M. and Tassoulas, J. L., "Dynamic Stiffness of Embedded Foundations," Recent Advances in Engineering Mechanics and Their Impact on Civil Engineering, ASCE, Vol. I, May, 1983, pp. 178-181.

2. Day, S. M. "Finite-Element Analysis of Seismic Scattering Problems," Ph.D. Thesis in Earth Sciences, University of California at San Diego, 1977.

3. Dobry, R., Oweis, I., and Urzua, A., "Simplified Procedures for Estimating The Fundamental Period of a Soil Profile," Bulletin of the Seismological Society of America," Vol. 66, No. 4, 1976, pp. 1293-1321.

4. Dobry, R., Vicente, E., O'Rourke, M. J., and Roesset, J. M., "Horizontal Stiffness and Damping of Single Piles," Journal of Geotechnical Engineering, ASCE, Vol. 108, No. GT3, Mar., 1982, pp. 439-459.

5. Dobry, R. and Gazetas, G., "Stiffness and Damping of Arbitrarily Shaped Machine Foundations," Journal of Geotechnical Engineering, ASCE, 1985, in press.

6. Dobry, R., Gazetas, G. and Stokoe, K. H., II, "Dynamic Response of Arbitrarily Shaped Foundations: Experimental Verification," Journal of Geotechnical Engineering, ASCE, 1985, in press.

7. Dominguez, J. and Roesset, J. M., " Dynamic Stiffness of Rectangular Foundations," Research Report R78-20, Dept. of Civil Engineering, M.I.T., 1978.

8. Erden, S. M. and Stokoe, K. H., II. "Effect of Embedment on Foundation Response," Research Report GR 85-8, The University of Texas at Austin, 1985

9. Gazetas, G., "Analysis of Machine Foundation Vibrations: State of the Art," Journal of Soil Dynamics and Earthquake Engineering, Vol. 2, No. 1, 1983, pp. 2-42.

10. Gazetas, G. "Rocking of Strip and Circular Footings," Proceedings of the International Symposium on Dynamic Soil-Structure Interaction/ Minneapolis/ September 4-5, 1984, 3-10, (Beskos, Krauthammer & Vardoulakis, editors).

11. Gazetas, G. and Roesset, J. M. "Forced Vibrations of Strip Footings on Layered Soils," Meth. Struct. Anal., ASCE, 1976, 1, 115.

12. Gazetas, G. and Dobry, R. "Horizontal Response of Piles in Layered Soils," Journal of Geotechnical Engineering, ASCE, Vol. 110, No. 1, January, 1984, pp. 20-40.

13. Gazetas, G. and Dobry, R., "Simple Radiation Damping Model for Piles and Footings," Journal of Engineering Mechanics, ASCE, Vol. 110, No. 6, June, 1984, pp. 931-956.

14. Gazetas, G., Dobry, R. and Tassoulas, J. L., "Vertical Response of Arbitrarily Shaped Embedded Foundations, " Journal of Geotechnical Engineering, ASCE, Vol. 111, No. 6, June, 1985, pp. 750-771.
15. Gazetas, G., "Vibrational Characteristics of Soil Deposits with Variable Wave Velocity," International Journal for Numerical and Analytical Methods in Geomechanics, Vol. 6, 1982, pp. 1-20.
16. Karasudhi, P., Keer, L.M. and Lee, S. L. "Vibratory Motion of a Body on an Elastic Half Plane," J. Appl. Mech., ASME, 1968, 35E, 697.
17. Kausel, E. "Forced Vibrations of Circular Foundations on Layered Media," Research Rep. R74-11, MIT, 1974.
18. Kaynia, A. M. and Kausel, E., "Dynamic Behavior of Pile Groups," Proceedings, Second International Conference on Numerical Methods in Offshore Piling, I.C.E., 1982.
19. Luco, J. E. and Westmann, R. A. "Dynamic Response of Circulatory Footings," J. Engng. Mech Divi., ASCE, 1971, 97, EM5, pp. 1381-1395.
20. Luco, J. E. and Westmann, R. A., "Dynamic Response of a Rigid Footing Bonded to An Elastic Half Space," Journal of Applied Mechanics, ASME, June, 1972, pp. 527-534.
21. Lysmer, J. "Lumped Mass Method for Rayleigh Waves," BSSA, 60, 1, February, 1970, pp. 89 -104.
22. Lysmer, J., "Foundation Vibrations with Soil Damping," Civil Engineering and Nuclear Power, ASCE, Vol. II, paper 10-4, 1980, pp. 1-18.
23. Lysmer, J. and Richart, F.E., Jr., "Dynamic Response of Footing to Vertical Loading," Journal of the Soil Mechanics and Foundations Division, ASCE, Vol. 92, No. SMI, 1966, pp 65-91.
24. Lysmer, J. and Kuhleymeyer, R. L. "Finite Dynamic Model for Infinite Media," J. Engrg. Mech Divi., ASCE, 1969, 95, EM4, 895.
25. Nogami, T., "Dynamic Group Effect in Axial Responses of Grouped Piles," Journal Geotechnical Engineering Division, ASCE, No. GT2, Vol. 109, 1983, pp. 228-243.
26. Novak, M. and Beredugo, Y. O., "The Effect of Embedment on Footing Vibrations", Proceedings First Canadian Conference on Earthquake Engineering Research, Vancouver, May, 1971.
27. Novak, M., Nogami, T., and Aboul-Ella, F., "Dynamic Soil Reactions for Plane Strain Case," Journal of Engineering Mechanics Division, ASCE, Vol. 104, No. EM4, Proc. Paper 13914, 1978, pp. 953-959.
28. Richart, F. E., Jr., Hall, J. R., Jr. and Woods, R. D., Vibrations of Soils and Foundations, Prentice-Hall International, Inc., New Jersey, 1970.
29. Roesset, J. M., "The Use of Simple Models in Soil Structure Interaction," Civil Engineering and Nuclear Power, ASCE, Vol. II, paper 10-3, 1980, pp. 1-25.
30. Roesset, J. M., "Stiffness and Damping Coefficients of Foundations," Dynamic Response of Pile Foundations, ASCE, 1-30, 1980.
31. Roesset, J. M., "Dynamic Stiffness of Pile Groups," ASCE Special Technical Publication on Analysis and Design of Pile Foundations, 1984, pp. 263-286.
32. Roesset, J. M. and Angelides, D., "Dynamic Stiffness of Piles," Numerical Methods in Offshore Piling, Institution of Civil Engineers, London, England, 1980, pp. 75-81.
33. Rücker, W. "Dynamic Behavior of Rigid Foundations of Arbitrary Shape on A Half Space," Earthquake Engineering and Structural Dynamics, Vol. 10, 1982, pp. 675-690.

34. Sheta, M. and Novak, M., "Vertical Vibration of Pile Groups," Journal of Geotechnical Engineering Division, ASCE, Vol. 108, No. GT4, 1982, pp. 570-590.
35. Thomas, G. E. "Equivalent Spring and Damping Coefficients for Piles Subjected to Vertical Dynamic Loads," Master of Engineering Thesis, Dept. of Civil Engineering, R. P. I., 1980.
36. Veletsos, A. S. and Wei, Y. T.,"Lateral and Rocking Vibration of Footings," J. Soil Mech. Found. Div., ASCE, 1971, 97, SM9, pp. 1227-1249.
37. Veletsos, A. J. and Verbic, B., "Basic Response Functions for Elastic Foundations, " Journal of the Engineering Mechanics Division, Vol. 100, No. EM2, 1974, pp. 189-201.
38. Veletsos, A. J. and Nair, V. V., "Torsional Vibration of Viscoelastic Foundations," Journal of the Geotechnical Engineering Division, ASCE, Vol. 100, No. GT3, 1974, pp. 225-246.
39. Whitman, R. V. and Richart, F. E., Jr., "Design Procedures for Dynamically Loaded Footings," Journal of the Soil Mechanics and Foundations Division, ASCE, Vol. 93, No. SM6, 1967, 169-193.
40. Wolf, J. and Von Arx, G. A., Impedance Function of a Group of Vertical Piles," Proceedings, ASCE Geotechnical Engineering Specialty Conference, Pasadena, 1978, pp. 1024-1041.
41. Wong, H. L and Luco, J. E. "Dynamic Response of Rigid Foundations of Arbitrary Shape," Earthquake Engineering and Structural Dynamics, Vol. 4, 1976, pp. 579-587.
42. Wong, H. L. and Luco, J. E., "Tables of Impedance Functions for Square Foundations on Layered Media," Soil Dynamics & Earthquake Engineering Vol. 4, 1985, pp. 64-81.

NUMERICAL MODELING FOR SOIL-FOUNDATION INTERACTION

Francisco Medina,[a] M. ASCE

A finite/infinite element technique is presented to solve three-dimensional elastic soil-foundation interaction problems in unbounded layered media. The technique is used to find the dynamic response of foundations subjected to harmonic loadings. By using finite and infinite elements, the size of the near field is kept small. Consequently, the system is characterized by relatively few degrees of freedom, thus providing the analyst with an inexpensive numerical solution.

INTRODUCTION

Field equations involving unbounded domains are encountered in many problems. Practical applications for such problems may be cited in various areas, including soil-foundation interaction. The soil media may be assumed to be semi-infinite in comparison to the foundation and the supported structure. This assumption ensures that under all circumstances there is no energy buildup in the considered domain. Therefore, the solution to soil-foundation interaction problems is sought in regions of small size compared with the surrounding media. The region where the solution is sought is commonly referred to as near field (nf). On the other hand, far field (ff) is the medium surrounding that region. When complex geometries and site conditions are considered, modeling of the semi-infinite continuum leads to discrete formulations based upon finite element techniques. Then, finite and infinite elements may be respectively used[2,9,3] to model the nf and ff. The interface between finite and infinite elements is for convenience defined as far-field boundary (ffb). Nevertheless, whereas finite elements are widely used as a means of discretization, approaches using infinite elements have only recently become available in the literature.[11,1]

By using finite and infinite elements together, the ff is represented in the analysis by means of deformable infinite elements, in the same sense that finite elements represent the nf. The formulation of this technique falls within the classical finite element method: discretize the continuum, assume shape functions, minimize errors by means of weighted residuals or variational principles, obtain a set of algebraic equations which characterizes the problem, etc. This technique therefore permits a direct one-step solution to the soil-foundation interaction problem, while preserving the flexibility of the classical finite element method.

INFINITE ELEMENTS FOR THREE-DIMENSIONAL ELASTIC WAVE PROPAGATION

Shape functions for multiwave propagation. The dynamic response of foundations embedded (or not) in semi-infinite media involves propagation of elastic waves in three dimensions and is extremely difficult to deal with. From known analytical closed form solutions[5] for semi-infinite media in harmonic free vibration, it becomes apparent that different types of waves propagate throughout the media, each disturbing every displacement component. Therefore, if meaningful infinite elements are sought to model the ff, different shape functions for each displacement should be developed. Figure 1 shows a typical discretization of an embedded foundation by means of finite and infinite elements. The displacements u^e for both types of elements are approximated by

[a]Facultad de Ciencias Físicas y Matemáticas, Universidad de Chile, Casilla 2777, Santiago.

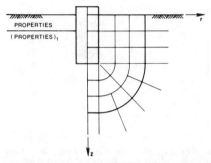

Figure 1.- Embedded foundation discretized with finite and infinite elements.

$$u^e(x) = N^e(x)r^e \tag{1}$$

where $N^e(x)$ contains the assumed shape functions at point x, and r^e contains the nodal displacements for elements e, with

$$N^e(x) = \begin{bmatrix} N^e_u(x) & 0 & 0 \\ 0 & N^e_v(x) & 0 \\ 0 & 0 & N^e_w(x) \end{bmatrix} \tag{2}$$

such that $N^e_y(x)$ contains the assumed shape functions at point x within element e, and corresponding to displacement component y ($y=u,v,w$). For finite elements

$$N^e_u(x) = N^e_v(x) = N^e_w(x) \tag{3}$$

For a typical displacement component y,

$$N_y(r,z) = f_y(r,z,k_R,k_S,k_P)F_y^{-1} \tag{4}$$

where F_y contains the nodal values of f_y, which in turn contains the wave components that are being transmitted (see Table 1). Details of the shape function generation may be seen elsewhere.[9] Then, by introducing the wave components into Eq.(4), the infinite element shape functions can be obtained. Obtaining these functions in closed form becomes extremely tedious; in practice, they are numerically computed. The infinite elements defined above are capable of transmitting Rayleigh, shear and compressional waves away from the nf. These elements approximately represent the dynamic stiffness of the ff for problems involving three-dimensional wave propagation.

Mapping functions. For the six-node axisymmetric infinite element shown in Fig.2, the interpolation function that maps the element from plane (r,z) onto plane (ξ,η), for node j, is

$$M_j(\xi,\eta) = L_j(\xi)L_j(\eta) \tag{5}$$

($0\leq\xi<\infty$), where L_j is the Lagrange polynomial for node j. The six-node axisymmetric infinite element shown in Fig.2 requires only three nodes to define its shape functions. The mapping functions for this element, however, is defined by all nodes. It is easily seen that this element may model layered unbounded media.

Shape functions for torsional wave propagation. Torsional vibrations are characterized by motions produced tangentially in planes perpendicular to the direction of wave propagation. Under appropriate definition of axes, the only

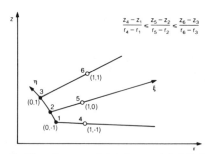

$$\frac{z_4 - z_1}{r_4 - r_1} \leqslant \frac{z_5 - z_2}{r_5 - r_2} \leqslant \frac{z_6 - z_3}{r_6 - r_3}$$

Figure 2.- Six-node parametric axisymmetric infinite element, with only three nodes having degrees of freedom (on $\xi=0$).

Table 1.- Displacement wave components for frequency dependent axisymmetric infinite elements.

DISPLACEMENT (y)	RAYLEIGH WAVE (f_y^R)	SHEAR WAVE (f_y^S)	COMPRESSIONAL WAVE (f_y^P)
u	$\left(e^{-p|z|} - \dfrac{2ps}{k_R^2 + s^2}e^{-s|z|}\right)e^{-(1+i)k_R r}$	$\dfrac{z}{R}e^{-(\xi + ik_S R)}$	$\dfrac{r}{R}e^{-(\xi + ik_P R)}$
v	$e^{-p|z|}\dfrac{e^{-(1+i)k_R r}}{k_R r}$	$e^{-(\xi + ik_S R)}$	$\hat{f}_v^R$
w	$\left(e^{-p|z|} - \dfrac{2k_R^2}{k_R^2 + s^2}e^{-s|z|}\right)e^{-(1+i)k_R r}$	$\dfrac{r}{R}e^{-(\xi + ik_S R)}$	$\dfrac{z}{R}e^{-(\xi + ik_P R)}$
v	$\hat{f}_v^R = e^{-s|z|}\dfrac{e^{-(1+i)k_R r}}{k_R r}$	-	-

- **PARAMETER DEFINITIONS:**

$r, z =$ cylindrical coordinates.

$R =$ $\sqrt{r^2 + z^2}$, radial coordinate.

$\xi =$ element parametric coordinate in the infinite direction.

$k_Q =$ wave propagation number ($Q=R$: Rayleigh; S: shear; P: compressional).

$s, p =$ $\sqrt{k_R^2 - k_{S,P}^2}$

nonvanishing displacement component is the tangential displacement v. Then, the shape function for node j and displacement v is[9]

$$N_j(\xi, \eta, k_S) = e^{-(1 + ik_S R_0)\xi}L_j(\eta) \tag{6}$$

where R_0 is a characteristic distance, say $R(\xi=0, \eta=0)$.

APPLICATION: DYNAMIC RESPONSE OF RIGID FOUNDATIONS

This problem reduces to that of finding the frequency-dependent foundation compliance matrix, $C(\omega)$.

Discretization. To take advantage of the method, the rigid foundations and the nf are modeled with a reasonably small number of nine-node finite elements, and the ff is modeled with even fewer six-node infinite elements. The foundations are chosen to be rigid only for comparison purposes. The response of the foundations is obtained for torsional (T), vertical (V), horizontal (H), and rocking (M) harmonic loadings. The nf characteristic matrices are computed using Gauss-Legendre quadrature with two integration points per direction. The ff dynamic stiffness is computed using reduced Gauss-Laguerre quadrature with two integration points in the finite direction and a larger but limited number of points in the infinite direction; this number varies with frequency, as indicated elsewhere.[9]

Numerical solutions are computed for discrete values of the dimensionless frequency

$$a_0 = a k_S = a\omega/V_S \tag{7}$$

where a is the foundation characteristic dimension and, k_S and V_S are the shear wave propagation number and velocity, respectively, of the semi-infinite medium. For a typical load J, the error (discrepancy) ϵ_J is defined as

$$\epsilon_J = \frac{\max_{a_0}\left\{|C_J(a_0)-\hat{C}_J(a_0)|\right\}}{|C_J(0)|} \tag{8}$$

where $C_J(a_0)$ and $\hat{C}_J(a_0)$ are respectively the exact (or any other available solution used for comparison) and the approximate numerical complex values, of the compliance diagonal term corresponding to load J. The global error in the foundation dynamic response is defined as

$$\epsilon = \left[\frac{1}{4}\sum_{J=T,V,H,M}\epsilon_J^2\right]^{1/2} \tag{9}$$

A summary of errors for the cases reported is shown in Table 2.

Rigid circular plate. The response of a rigid circular plate on the surface of a half-space was obtained for harmonic loadings. The nf is modeled by nineteen axisymmetric finite elements and the ff by four infinite axisymmetric elements, as

Table 2.- Compliance function errors for rigid foundations on semi-infinite media.

LOADING	RIGID HEMISPHERE			RIGID PLATE			RIGID PLATE ON LAYER OVER HALF-SPACE	
	DOF[†]	ERROR (%)		DOF[†]	ERROR (%)		DOF[†]	ERROR (%)
		$R^*/R_0=2.00$	$R^*/R_0=2.25$		$R^*/R_0=2.00$	$R^*/R_0=2.25$		$R^*/R_0=2.25$
TORSIONAL	24	2.5	2.5	87	4.2	4.1	82	-
VERTICAL	44	13.3	6.7	146	6.6	7.3	172	6.5
HORIZONTAL	80	13.9	12.2	225	7.9	6.5	298	10.0
ROCKING	80	6.4	4.3	225	6.9	6.7	268	12.8
GLOBAL ERROR		10.2	7.9		7.7	6.9		10.1

†Number of degrees of freedom.

shown in Fig.3a. The ffb is located two plates radii away from the plate origin. It was observed that by modifying the ratio R^*/R_0 (see Fig.3a), a better solution may be obtained. For a ratio of 2.25, the global error is found to be minimum. The compliance functions computed, shown with dots in Figs.3b-f, are compared with the corresponding exact solutions,[6,12] which are shown with solid lines. It is observed that the computed numerical solutions are in good agreement with the exact solutions.

Rigid hemispherical body. The response of a rigid hemispherical body embedded in a semi-infinite medium was obtained for harmonic loadings. The nf is modeled by eight axisymmetric finite elements and the ff by four infinite axisymmetric elements, as shown in Fig.4a. The ffb is located one and one-half hemisphere radii away from the hemisphere origin. In this case, the error shows similar behavior to that found previously when the ratio R^*/R_0 varies. The minimum error is also found in the neighborhood of 2.25. The compliance functions computed, shown with dots in Figs.4b-f, are compared with other available solutions,[7,4] which are shown with solid lines. Despite the few degrees of freedom used for discretization, there is good agreement between both sets of solutions.

Rigid plate on a single layer over a half-space. The response of a rigid circular plate on the surface of a homogeneous, isotropic single layer resting over a homogeneous, isotropic semi-infinite medium was obtained for harmonic loadings. The thickness of the layer is 0.2 times the plate radius. The nf is modeled by twenty-four finite elements and the ff by five infinite elements, as shown in Fig.5a. The ffb is approximately located at two plate radii away from the plate origin. The compliance functions computed are shown with dots in Figs.5b-f, where the exact solutions[6] are shown with solid lines for comparison.

General remarks. In spite of the few degrees of freedom used for discretization it is observed that the computed numerical solutions are generally in good agreement with the solutions shown. Maximum discrepancies are below 10.5%, see Table 2. Since the numerical finite element approximation for multiwave propagation is not defined for zero frequency, the compliance functions shown for zero frecuency were obtained with shape functions simpler than those outlined previously, but developed[10] under the same basic principles.

CONCLUSIONS

A direct finite/infinite element method for the linear elastic analysis of dynamic soil-foundation interaction has been briefly outlined. By modeling the far field with infinite elements, this method permits a significant reduction in the size of the near field. Solutions were carried out with a small number of degrees of freedom, which provided inexpensive answers and satisfactory accuracy. With this method, the analysis of layered systems, foundation embedment or rigid bases is easily undertaken, involving no extra effort.

ACKNOWLEDGEMENTS

This work was partially supported by the Chilean National Council for Science and Technology (CONICYT) and URS/John A. Blume & Associates, Engineers.

REFERENCES

1. Bettess, P., 'Infinite Elements,' *Int.J.num.Meth.Engng.*, 11,53-64(1977).

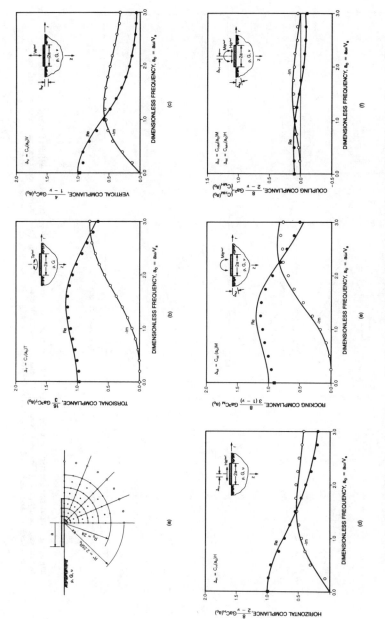

Figure 3.- Rigid circular plate on a homogeneous, isotropic half-space ($\mu=1/3$) subjected to harmonic loading. (a) Element mesh. (b) Torsional compliance function. (c) Vertical compliance function. (d) Horizontal compliance function. (e) Rocking compliance function. (f) Coupling compliance function.

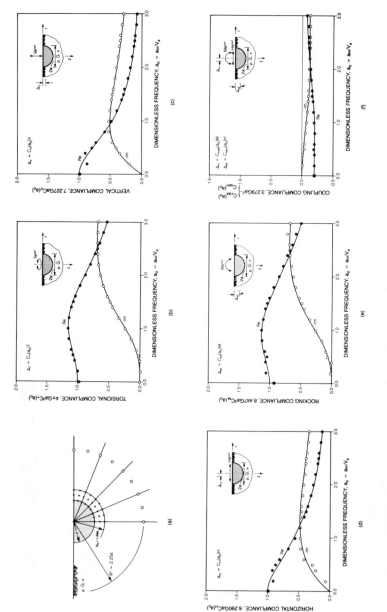

Figure 4.– Rigid hemispherical body embedded in a homogeneous, isotropic semi-infinite medium ($\nu = 1/4$) subjected to harmonic loading. (a) Element mesh. (b) Torsional compliance function. (c) Vertical compliance function. (d) Horizontal compliance function. (e) Rocking compliance function. (f) Coupling compliance function.

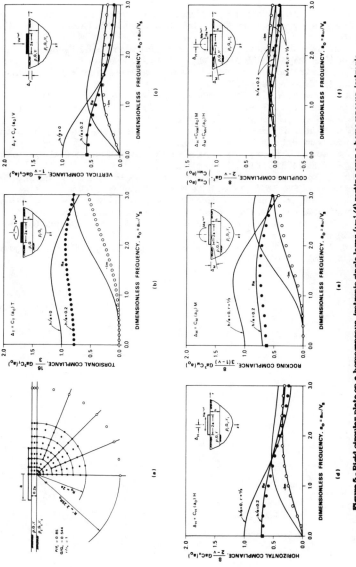

Figure 5.- Rigid circular plate on a homogeneous, isotropic single layer ($\nu=1/4$) resting over a homogeneous, isotropic half-space ($\nu=1/4$) subjected to harmonic loading. (a) Element mesh. (b) Torsional compliance function. (c) Vertical compliance function. (d) Horizontal compliance function. (e) Rocking compliance function. (f) Coupling compliance function. (The compliance functions for a rigid circular plate on a homogeneous, isotropic half-space are shown for reference).

2. Bettess, P. and O.C. Zienkiewicz, 'Diffraction and Refraction of Surface Waves Using Finite and Infinite Elements,' *Int.J.num.Meth.Engng.*, 11,1271-1290 (1977).

3. Chow, Y.K. and I.M. Smith, 'Static and Periodic Infinite Solid Elements,' *Int.J. num.Meth.Engng.*, 17,503-526(1981).

4. Day, S.M., 'Finite Element Analysis of Seismic Scattering Problems,' *Ph.D.thesis*, U. California, San Diego (1977).

5. Graff, K.F., *Wave Motion in Elastic Solids*, Ohio State Univ. Press (1975).

6. Luco, J.E., 'Impedance Functions for a Rigid Foundation on a Layered Media,' *Nuclear Engng.&Design*, 31,204,217(1974).

7. Luco, J.E., 'Torsional Response of Structures for SH Waves: The Case of Hemispherical Foundations,' *BSSA*, 66,109-123(1976).

8. Luco, J.E. and R.A. Westmann, 'Dynamic Response of Circular Footings,' *proc. ASCE*, 97,EM5,1381-1395(1971).

9. Medina, F., 'Modelling of Soil-Structure Interaction by Finite and Infinite Elements,' *rep.UCB/EERC-80/43*, U. California, Berkeley (1980).

10. Medina, F., 'Infinite Elements for Modeling Layered Soil-Structure Interaction,' submitted to *Earthqu.Eng.Struct.Dyn.*, for possible publication (1985).

11. Ungless, R.F., 'An Infinite Finite Element,' *M.A.Sc.thesis*, U. British Columbia (1973).

12. Veletsos, A.S. and Y.T. Wei, 'Lateral and Rocking Vibration of Footings,' *proc. ASCE*, 97,SM9,1227-1248(1971).

LOW TUNED COMPRESSOR FOUNDATIONS ON SOFT CLAY

by Christian Madshus*, Farrokh Nadim*, Arne Engen*,
and Arne Martin Lerstøl**

A B S T R A C T

This paper describes the design and testing of two low-tuned foundations for piston compressors at a site with soft, sensitive clay,
and about 40 m depth to bedrock. Each of the two resilient foundation
blocks are supported on four end bearing piles. The piles have a special design which serves two purposes: they act as supports for the
static loads and as axial spring elements to gain the desired natural
frequency of the foundations. Results of a numerical procedure developed for evaluation of the maximum transient response due to passing
through the resonance frequency of the foundations during starting and
stopping of the compressors are presented. The paper describes how
vibration measurement techniques were used during test running of the
compressors to detect points of undesired contact between the resilient foundation blocks and the outer foundations.

1. INTRODUCTION

Foundations for two compressors were to be designed for one of Norsk
Hydro's petrochemical plants in Norway. Both compressors were of the
piston type, with low operating speeds and significant vertical unbalanced forces.

Figure 1 presents a plan of the site, with the locations of the two
piston compressors. It was necessary to take special care in design
of the foundations due to the poor soil conditions and the presence of
vibration-sensitive installations nearby. Vibration measurements
during testing of the compressors indicated unexpected and unsatisfac-
tory behaviour of the foundations, and corrective measures were taken
to establish the conditions assumed in design.

---

* Research engineers, Norwegian Geotechnical Institute, Oslo, Norway
** Consulting engineer, Torgersrud og Lerstøl A/S, Oslo, Norway

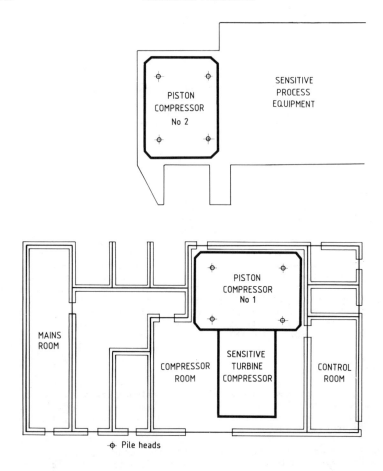

Fig 1   Plan of site and location of compressor

2.   GEOTECHNICAL FOUNDATION PROBLEMS

The soil at the site consists of 4 m fill overlaying 5 m of fine sand
-coarse silt. Below the sand layer, there is a layer of clayey silt -
silty clay which goes down to approximately 30 m below the ground
level. Underneath the clay layer, there is gravel which extends down
to the bedrock at 35-45 m depth.

In the silty clay, the undrained shear strength measured by laboratory
fall cone and by unconfined compression tests, increases from approximately 40 kPa at 15 m depth to 60 kPa at 25 m depth, giving an s_u/p_o'-
ratio of 0.22-0.23. The sensitivity of this clay is between 10 and 20.

Figure 2 presents a typical soil profile at the site. Rotation
sounding measurements shown in this figure give qualitative information about soil strength and stiffness (Hvorslev, 1949)

The area has been filled up during two periods. About 25 years ago, a
2.5 m thick layer of rockfill was placed at the site. Ten years later,
1.5 m of organic fine sand was placed on top of the previous fill.
The fill has given rise to long term settlements in the area that are
presently (1984) occuring at a rate of about 5 mm/year. As a consequence of these rather significant settlements, all sensitive buildings
must be founded on end-bearing piles to bedrock. Experience has shown
that the ongoing settlements cause negative skin friction, which gradually may lead to serious overloading of this type of piles. To avoid
this problem, the piles were coated by bitumen before installation.

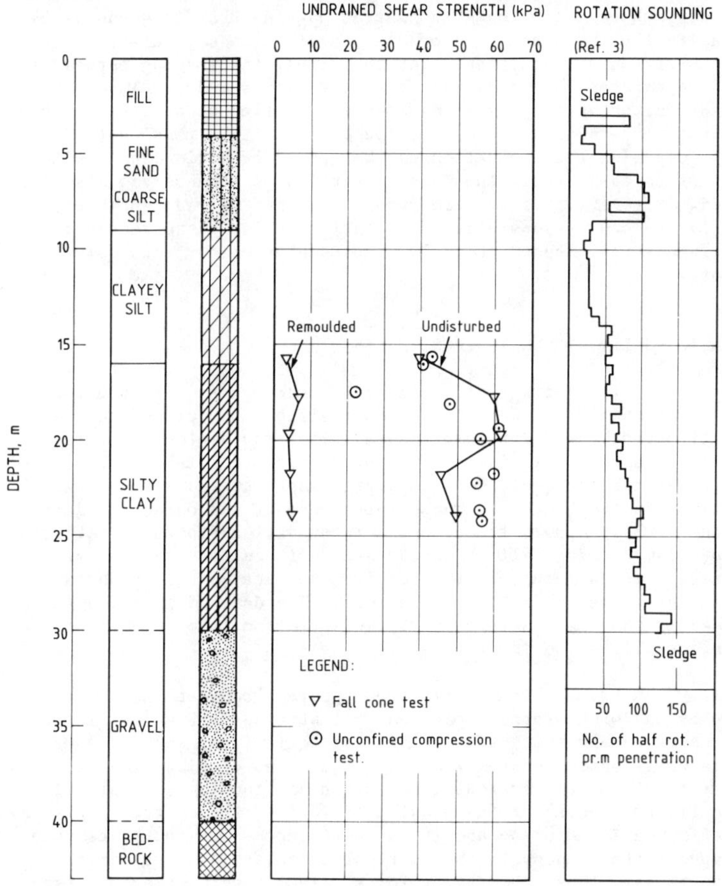

Fig. 2   Typical soil profile at the site

3. DYNAMIC FOUNDATION PROBLEMS

Compressor No. 1 is a twin cylinder unit with an operating speed of
735 RPM. Compressor No. 2 has one cylinder and operates at 520 RPM.
The primary vertical unbalanced forces are about 70 kN and 25 kN,
respectively. More details are given in Section 5.

The main criteria for the foundation design were to keep the vibrations at the support points of the compressors, and at the adjacent
sensitive installations, below specified limits. Due to the poor soil
conditions and the ongoing settlements, the foundations had to be supported on end-bearing piles driven to bedrock.

Both high tuned and low tuned foundation solutions were evaluated.
High tuned foundations have natural frequencies above the excitation
frequency, whereas low tuned foundatios have natural frequencies below
the excitation frequency. In order to avoid excessive disturbance of
the sensitive clay when installing the piles, the spacing between them
had to be greater than a certain distance. This made it impossible to
gain sufficient vertical stiffness from the piles alone for a high
tuned solution and still keep the foundation within reasonable dimensions. Coupling the foundation to the ground by adding short bitumen
coated piles to increase the dynamic vertical stiffness was also considered. However, this solution would have led to unacceptable vibrations in the nearby sensitive installations. For these reasons, low
tuned foundations seemed the only viable solution for both compressors.

4. BASIC DESIGN AND DESIGN PROCEDURES

In agreement with Norsk Hydro, a somewhat unconventional solution,
illustrated in Fig. 3, was chosen for resilient support of the heavy
foundation blocks for the compressors. Each foundation block rests on
four piles, each consisting of two concentric steel tubes. The closed
ended outer tube is equipped with a pile shoe, coated with bitumen,
and driven to the bedrock. These tubes support the outer foundation.
The inner tubes support the compressor foundation blocks. Their free
length, diameter and wall thickness are designed to obtain axial
stiffnesses which result in the required vertical and rocking natural
frequencies for each of the foundations. The desired free lengths are
obtained by pouring concrete in the lower part of the outer tubes
before installing the inner ones.

Laterally, the foundation blocks are supported against the outer foundation by sandwich neoprene pads, as illustrated in Fig. 3. These
pads are extremely stiff in the axial direction, but soft in shear.
This lateral support system restricts the modes of vibration of the
foundation blocks to vertical translation and rocking. By aligning
vertically the centre of the unbalanced forces, the centre of gravity
of the foundation blocks, and the support centre for the piles, one
may approximately uncouple the vertical translation modes and the
rocking modes. Thus, the theory for single degree-of-freedom systems
may be used for modelling each of the modes separately.

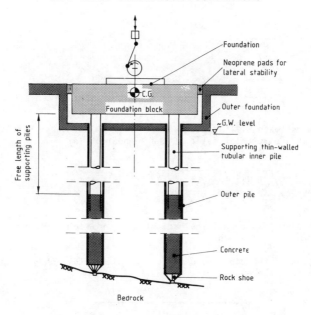

<u>Fig. 3</u>   Resilient foundation supported on thin walled piles –
          Basic design principles

4.1 Steady state conditions

Since the translational and rocking modes can be assumed seperable,
the vertical steady state vibration response at any point of the foundation may be evaluated from the following equations (Crede, 1965):

$$x_{i_t} = \frac{F_{ei}}{4k \sqrt{\left[1-(\frac{\Omega i}{\omega_{n_t}})^2 \right]^2 + \left[2\zeta_t(\frac{\Omega i}{\omega_{n_t}}) \right]^2}} \tag{1}$$

for vertical vibration, and

$$x_{i_r} = \frac{M_{ei} \cdot d}{kD^2 \sqrt{\left[1-(\frac{\Omega i}{\omega_{n_r}})^2 \right]^2 + \left[2\zeta_r(\frac{\Omega i}{\omega_{n_r}}) \right]^2}} \tag{2}$$

for rocking.

Here, the index i refers to the harmonic order number.

The natural angular frequencies for vertical translation and rocking
are:

$$\omega_{n_t} = \sqrt{\frac{4k}{M}} \tag{3}$$

$$\omega_{n_r} = \sqrt{\frac{kD^2}{I}} \tag{4}$$

A maximum estimate of the root-mean-square (RMS) of the vibration velocity at a particular point of the foundation is:

$$\dot{x}_{RMS} = \sqrt{\sum_i \frac{\Omega_i^2 \; (x^2 i_t + x^2 i_r)}{2}} \tag{5}$$

where the sum includes all the harmonic orders which contribute significantly to the vibration level.

The fraction of the unbalanced force transmitted to each pile can be estimated from the equation:

$$F_{t_i} = F_{e_i} \sqrt{\frac{1 + [2\zeta_t(\frac{\Omega_i}{\omega_{n_t}})]^2}{[1-(\frac{\Omega_i}{\omega_{n_t}})^2]^2 + [2\zeta_t(\frac{\Omega_i}{\omega_{n_t}})]^2}} \tag{6}$$

where F_t represents the amplitude of the transmitted force.

4.2 Passing through resonance

One of the problems with low tuned foundations is that the excitation will pass through their resonance conditions when the compressors are starting up or shutting down. Experience has shown that these conditions can result in large transient vibrations (Crede, 1965). However, no straight-forward procedure seems to exist for estimating these vibrations, and an approach, based on a numerical solution of the governing equations, was developed as a part of this project.

The governing equation of motion for a machine with rotating a eccentric mass is:

$$\ddot{x} + 2\zeta\omega_n\dot{x} + \omega_n^2 x = \frac{m_e e}{M} \cdot \Omega^2(t) \cdot \sin\Omega(t)t \tag{7}$$

For definition of terms refer to Appendix II.

A frequently made assumption when solving Eq. (7) is that $\Omega(t)$ is constant. This assumption implies that the onset of excitation is sudden. For a constant $\Omega(t) = \Omega$, and at-rest initial conditions, the solution to Eq. (7) is:

$$x = \frac{m_e e \cdot \Omega^2}{M \cdot \omega_n^2} \cdot \frac{A_1 \sin\Omega t + A_2 \cos\Omega t + e^{-\zeta\omega_n t}[A_3 \sin\omega_d t - A_2 \cos\omega_d t]}{A_1^2 + A_2^2} \qquad (8)$$

where

$$A_1 = 1 - \left(\frac{\Omega}{\omega_n}\right)^2$$

$$A_2 = -2\zeta \frac{\Omega}{\omega_n}$$

$$A_3 = \frac{\Omega}{\omega_d}\left[2\zeta^2 - 1 + \left(\frac{\Omega}{\omega_n}\right)^2\right]$$

$$\omega_d = \omega_n \sqrt{1-\zeta^2} = \text{damped natural frequency of the system}$$

Although the free vibration part of this solution (terms with ω_d) is damped out quickly, it plays an important role in the maximum transient response of the system.

Equation (8) is valid when there is no transition time during which the driving frequency reaches its steady-state value from an initial value of zero. In reality, however, this transition time is never zero. The exact variation of $\Omega(t)$ between zero and its steady-state value, Ω_1, depends mostly on the nature of the torque that rotates the machinery. For example, if a constant torque is applied to the shaft of the machine, $\Omega(t)$ would have a more or less linear variation in the transition period:

$$\Omega(t) = \Omega_1 \frac{t}{T_0} \qquad t < T_0$$

$$\Omega(t) = \Omega_1 \qquad t > T_0 \qquad (9)$$

where T_0 = time required to reach the steady-state operating frequency

If, on the other hand, the torque that rotates the machine comes from an asynchronous electric motor, $\Omega(t)$ could be approximated by an exponential variation during the start-up transition period. The results presented in this paper are based on variation of $\Omega(t)$ according to Eq. (9).

In general, when $\Omega(t)$ varies with time, it is not possible to find a closed form solution to Eq. (7). Thus, a numerical solution is attempted. The explicit central difference method is used in this study. The following finite difference approximations for acceleration and velocity are used in the central difference method:

$$\ddot{x}_t = \frac{1}{\Delta t^2} (x_{t-\Delta t} - 2x_t + x_{t+\Delta t})$$

$$\dot{x}_t = \frac{1}{2\Delta t} (- x_{t-\Delta t} + x_{t+\Delta t})$$

(10)

where Δt = time increment used in the numerical analysis.

Detailed convergence and stability analysis for the central difference method is carried out by Bathe (1982) for linear systems. In this study, a value of $\Delta t = 0.01T_1$, where T_1 is the steady-state operational period of the rotating machinery (i.e. $T_1 = 2\pi/\Omega_1$), has been used. Note that T_1 is smaller than the natural period of the system for low-tuned foundations.

The numerical solution is compared to the closed form solution for the case of the sudden onset of excitation (Eq. 8) in Fig. 4. The system has a natural period of $T = 1$ and a damping ratio of $\zeta = 2\%$. The numerical and the closed form solutions are indentical to within the thickness of the curves in Fig. 4, giving confidence in the numerical solution.

This numerical solution is used to study the solution of Eq. (7) when the driving frequency $\Omega(t)$ varies according to Eq. (9). Only situations with $\Omega_1 \gg \omega_n$, where the problem of passing through resonance may be important, have been considered. A typical numerical solution is shown in Fig. 5. Displacements are normalized with respect to $m_e e/M$, and time is normalized with respect to the natural period of the system $T = 2\pi/\omega_n$. The results shown in Fig. 5 are for $\Omega_1 = 2\omega_n$, $T_0 = 50T$ in Eq. (9), and $\zeta = 4\%$.

The response shown in Fig. 5 has several interesting features. First of all, two types of transients are observed. The first transient occurs when time is less than T_0 (around TIME = 13). This transient is associated with passing through resonance. The second transient occurs right after the operational frequency reaches its steady-state value at time T_0, and is very similar to what happens when the onset of excitation is sudden (Fig. 4c). The transient at time T_0 is caused by the abrupt change of the time history of $\Omega(t)$ at T_0 (discontinuous $d\Omega/dt$). In real life situations, however, $\Omega(t)$ has a smooth transition to its steady-state value. The transient at time T_0 was not observed when variations of $\Omega(t)$ with a smooth transition (continuous $d\Omega/dt$) at T_0, were considered. Thus, this transient is only due to the simplifying assumptions made for the analysis and has no physical significance.

The second interesting feature of the response in Fig. 5 is that the vibration frequency of the foundation at times just before T_0 appears to be equal to $2\Omega_1$, and that the maximum response at times before T_0 occurs when the driving frequency is less than the natural frequeny of the system. To examine these features in more detail, one should take a closer look at Eq. (7). The right hand side of Eq. (7) has a sinusoidal variation of the form $\sin(\Omega_1/T_0)t^2$. At a given value of time,

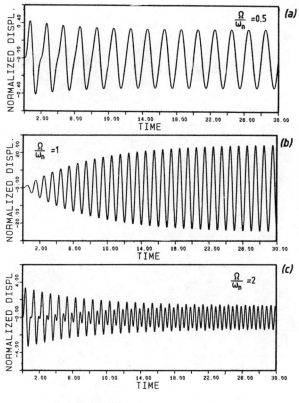

―――― *Analytical solution*
― ― ― *Numerical solution*
Note: Both solutions are identical to within the thickness of curves.

Fig. 4   Comparison of numerical solution and closed form solution for
          sudden onset of excitation, ζ = 2%, T = 1

$t = t^* < T_o$, this term can be linearised by the Taylor expansion to
read:

$$\sin \frac{\Omega_1}{T_o} t^2 \approx \sin \left[\frac{2\Omega_1}{T_o} t^* t + \frac{\Omega_1}{T_o} t^{*2} \right] \tag{11}$$

Thus the <u>apparent frequency</u> of excitation at time t^* is $2(\Omega_1/T_o)t^*$ or
$2\Omega(t)$! This means that passing through resonance occurs when $\Omega(t) \approx$
$0.5\omega_n$, and that the apparent frequency of loading just before T_o is
equal to $2\Omega_1$. These conditions are entirely consistent with the
results shown in Fig. 5.

The maximum transient displacement during the process of passing
through resonance (t<To), is presented in normalized form in Fig. 6

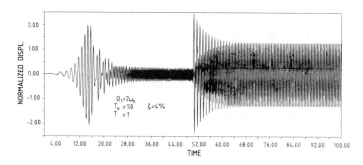

Fig. 5   Linear increase of $\Omega(t)$ from zero to Ω_1 during $T_0 = 50$ T

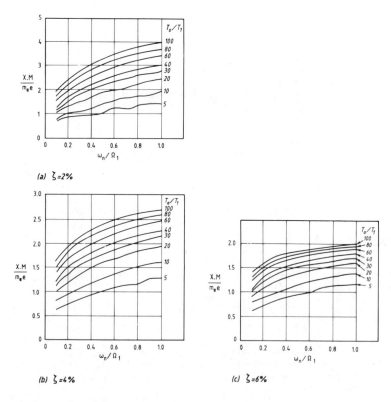

Fig. 6   Normalized maximum transient displacement while passing through resonance

for different damping ratios. A few parametric studies where the
variation of $\Omega(t)$ with time was different from Eq. (9) were performed.
The parametric studies showed that the maximum transient displacement
is not very sensitive to the exact variation of $\Omega(t)$. Thus, the curves presented in Fig. 6 can be used for design, even if $\Omega(t)$ does not
exactly vary according to Eq. (9). The maximum transient velocity and
the maximum transient acceleration can be estimated by multiplying the
maximum transient displacement by ω_n and ω_n^2, respectively.

5. SPECIFIC DESIGN OF THE COMPRESSOR FOUNDATIONS

The operating speeds, excitation frequencies and the unbalanced force
and moment amplitudes for the two compressors are listed in Table 1.

Table: 1   Operating speeds and unbalanced forces for the
compressors.

COMPRESSOR UNIT.	OPERATING SPEED	VERTICAL UNBALANCED FORCES		UNBALANCED TORQUE
		1. Harm kN	2.Harm kN	1. Harm kNm
NO 1	735 RPM 12.3 Hz	70.5	0.6	26.4
NO 2	520 RPM 8.7 Hz	25.0	5.0	~0

The main criteria for the foundation design were to keep the vibration levels below specific limits both on the foundation itself and on
the adjacent sensitive installations. The limits were specified by
the suppliers of the compressors, with reference to the German
industrial standard VDI 2056 (Verein Deutscher Ingenieure, 1964).

- In continuous operation, the vibration velocity level at the
  compressor support points should not exceed 4.0 mm/s RMS.

- Anywhere else on the foundation blocks, the vibration level should
  not exceed 5.0 mm/s RMS.

- On the nearby sensitive installations, the vibration level should
  be limited to 0.5 mm/s RMS.

- During start-up and shut-down of the compressors, the above limits
  could be exceeded by a factor of 2.0.

To design for the steady-state situation, equations 1 through 6 were
used, assuming a damping ratio of ζ = 5% for all modes.

For the situation of passing through resonance during starting and
stopping, the normalized diagrams in Fig. 6 were applied, assuming the
time to reach steady state operating speed, T_0, was 5 seconds, and
ζ = 5%. Based on these assumptions, the transient vibrations during
starting and stopping were not critical in the design of the foundations.

The calculated static response was also checked based on the criterion
that a 100 kN dead weight placed in the most unfavourable position,
should not lead to a tilt of the foundation exceeding 1/300. This
criterion was found not to be critical.

Table 2 lists the masses, dimensions, and the required spring constants, resulting from this design.

Table:2   Key figures for the resilient foundations.

COMPRESSOR UNIT.	PILES	RESILIENT FOUNDATION BLOCKS			
	VERTICAL SPRING CONSTANTS (4 piles) kN/m	MASS INCL. COMP. kg	DIMENSIONS LxWxT m	REQUIRED MASS DENSITY kg/m^3	MOM. OF INERTIA INCL. COMP. kgm^2
NO 1	210 000	300 000	9.4x7.3x1.5	2 820	1.4 · 10^6
NO 2	152 000	170 000	8.1x5.6x1.5	2 380	0.32 · 10^6

The design specified in Table 2 provides a safety factor of 1.5
against exceeding the specified vibration limits.

For compressor unit No. 2, ordinary reinforced concrete could be used
in the foundation block. For unit No. 1, however, heavy aggregates
had to be added to the concrete to obtain the required mass.

Table 3 compares the compressor excitation frequencies with the computed vertical natural frequencies for the two compressors. For unit
No. 1, the ratio between the two is 2.4, which is safisfactory. For
unit No. 2, the ratio is only 1.8, which is marginal.

Table: 3   Excitation and natural frequencies assumed
in design.

COMPRESSOR UNIT.	EXCITATION FREQUENCY f_e(Hz)	NATURAL FREQUENCY f_n(Hz)	FREQUENCY RATIO f_e/f_n
NO 1	12.3	5.1	2.4
NO 2	8.7	4.8	1.8

The natural frequencies of the rocking modes are lower than those of
vertical translation and are not critical.

The free length, diameter and wall thickness of the inner piles were
designed to obtain the required axial spring constants (Table 2).
Furthermore, the piles were designed to have sufficient safety against
buckling. Table 4 lists the dimensions of the piles for the two foundations, as assumed in the design. For the concrete in the bottom of
the piles a dynamic modulus of elasticity of 3.0 · 10^7 kPa was
assumed.

Table: 4   Key figures for the thin walled supporting piles.

COMPRESSOR UNIT.	FREE LENGTH m	DIAMETER m	WALL THICKNESS mm
NO 1	19.0	0.335	6.0
NO 2	21.0	0.335	3.6

Figure 7 shows the details of the double pile construction.

Reinforced concrete is poured in the bottom part of the outer piles to obtain the desired length of the supporting inner piles.

The space between the inner and the outer piles and the interior of the inner piles is filled with water and inhibitor to prevent corrosion. The water filling also adds damping to the lateral, flexural

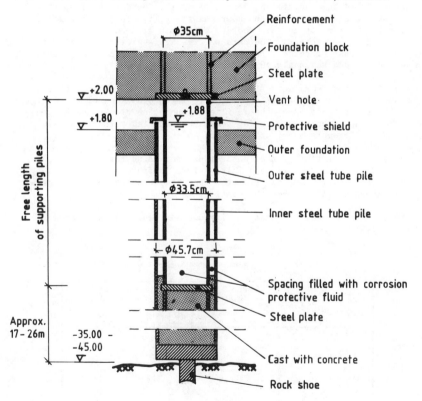

Fig. 7   Supporting piles - Design details

vibrations, if some of these modes should happen to be excited by the
compressors.

If the inner piles are entirely filled, the nearly incompressible water
will add substantial undesired axial stiffness to the piles. To
ensure that this is not happening, vent holes are provided near the
top of the inner piles.

A shield is supplied to prevent foreign matter from accidentally
entering the space between the inner and the outer piles.

6. VIBRATION MEASUREMENTS DURING TEST RUNNING OF THE COMPRESSORS

A measuring program was planned and executed to monitor the foundation vibrations during test running of the compressors.

For the measurements, six servo accelerometers were used for simultaneous recording at different locations on the foundation slab, at
nearby structures, and at the compressor itself. The accelerometer
signals were amplified and recorded both on a strip chart recorder and
on an FM-tape recorder. The data were mainly evaluated manually from
the strip chart recorder, but some sequences of particular interest
were digitised and analysed by Fourier transformation techniques.

When the compressors were started up, it soon became clear that the
foundations did not behave according to the proceedings of the design
analyses. Vibrations at the resilient foundation blocks and at the
surrounding structures were far above the expected values, and a
period of "trouble shooting" started.

By impact excitation of the foundations, one found that their natural
frequencies were too high and fairly close to frequencies excited by
the compressors. Damping estimated from the logarithmic decrement was
also substantially higher than anticipated.

The vibration levels and natural frequencies measured are listed in
Table 5.

Table:5   Measured foundation vibration levels and natural frequencies at different stages.

STATUS OF FOUNDATION.	COMPRESSOR UNIT NO 1 VIBRATION LEVEL.			COMPRESSOR UNIT NO 2 VIBRATION LEVEL.		
	MEAN mm/s_{RMS}	MAX. mm/s_{RMS}	NATURAL FREQU. Hz	MEAN mm/s_{RMS}	MAX. mm/s_{RMS}	NATURAL FREQU. Hz
At initial start up	6.5	8.0	8.6	-	-	-
After first cleaning of joints.	5.0	11.7	8.4	7.6	13.2	7.6
After final cleaning of joints.	4.6	8.5	~8.1	6.6	12.5	~6.8
After remooving contacts at pile heads.	2.6	3.6	5.7	3.6	5.0	5.9

The mean vibration levels represent the average of levels measured at each point on the foundation. The maximum levels were usually found close to the foundation edges and indicate substantial rocking motions.

These findings indicated undesired mechanical contact between the resilient foundation blocks and the outer foundations shown in Fig. 3.

By measuring the vibrations along the foundation periphery at both sides of the joint separating the resilient foundation blocks from the outer foundation, one localized the regions of major contact. An example of such measurements is presented in Fig. 8, which shows the results of the measurements of compressor unit No. 1. By visual inspection, structural elements attached to both the resilient and the outer part of the foundation were found. It was also found that the polystyrene foam, used as formwork when pouring the foundations, still remained in the joint. After removing these contacts, the situation was improved, but it was still not satisfactory, as can be seen from Table 5.

During a more thorough inspection, some concrete was discovered in the joints, and this was removed. This further improved the situation. However, after this final cleaning of the joints, the foundations did still not behave according to the design analyses, as can be seen from Fig. 8 and Table 5.

From the measurements, the modes of vibration of the foundations could be determined, as illustrated in Fig. 9. The vibration modes were dominated by rocking motion along diagonal lines, indicating some degree of fixity at one or two pile heads. As shown in the figure, a mode could change from one day to another, indicating that the support conditions were not stable.

By inspecting the pile heads it was found that the clearance between the protective shields, shown in Fig. 7, and the outer piles had been insufficient to accommodate the static deflection of the inner piles. Thus, contact had developed. As can be seen from Fig. 8 and Table 5, after these final improvements, both the natural frequencies and vibration levels dropped to values close to those predicted by the design analyses. In this final condition, the vibration modes were purely vertical translation without any rocking.

Table 6 compares the measured natural frequencies and vibration levels to those predicted in the design.

Table: 6   Comparison of predicted and measured natural frequencies and vibration levels.

COMPRESSOR UNIT.	NATURAL FREQUENCY		VIBRATION LEVEL			
	PREDICTED IN DESIGN Hz	MEASURED Hz	SPECIFIED MAX. LIMIT mm/s$_{RMS}$	PREDICTED IN DESIGN mm/s$_{RMS}$	COMPUTED ACC.TO NAT.FR. mm/s$_{RMS}$*	MEASURED mm/s$_{RMS}$
No. 1	5.1	5.7	4.0	2.6	2.7	2.6
No. 2	4.8	5.9	4.0	2.7	3.5	3.6

*) Assuming that the difference between designed and measured natural frequency is due to pile stiffness alone.

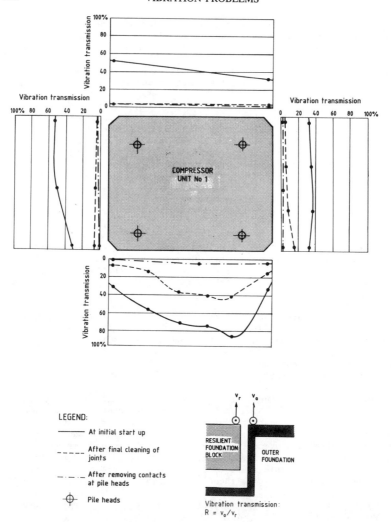

<u>Fig. 8</u>   Vibration transmission from the resilient foundation to outer
foundation – Compressor unit No. 1

As can be seen, the measured natural frequencies are higher than
predicted, especially for compressor unit No. 2.

For unit No. 1 the vibration levels measured agree well with those predicted in the design, but for unit No. 2 the measured values are significantly higher. However, by assuming that the higher natural frequencies were due to stiffer piles, the new vibration levels were com-

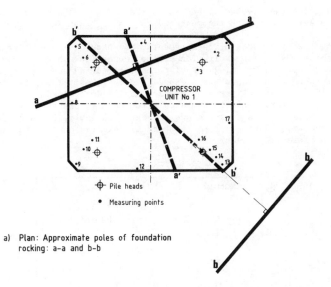

a) Plan: Approximate poles of foundation
   rocking: a–a and b–b

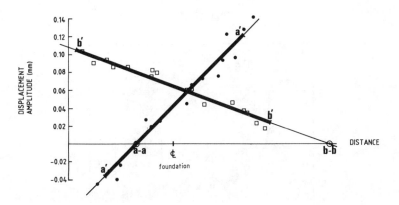

b) Measured values projected on the lines a' – a' and b' – b'

Fig. 9   Rocking performance of resilient foundation due to partial
          contact at pile heads – Change in conditions from one day to
          another – Compressore unit No. 1

pared using the same equations as in the design. These backcalculated values agree well with the measured values for both foundations.

During the first sets of measurements, the natural frequencies were determined by impact excitation of the foundations when the compressors were not running. During the later measurements, the compressors could no longer be stopped. However, as illustrated in Fig. 10, the natural frequencies were easily determined from the power spectra of vibration records taken with running compressors. The natural frequencies measured are presented in Tables 5 and 6. From the spectra, the response to the harmonic excitation and its higher harmonics is very clear.

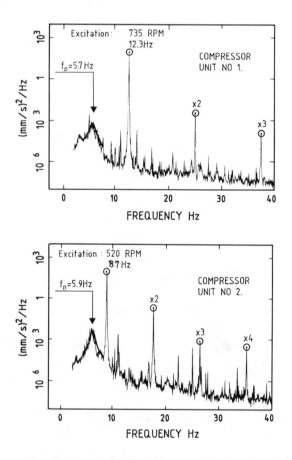

Fig. 10   Power spectra of vibrations at foundation - After removing
           contacts at pile heads

During the earlier phase of the test running process, vibrations were
recorded during starting and stopping of the compressors. In Fig. 11
one such trace is compared to a corresponding trace predicted by the
numerical model described in Section 4. Qualitatively, the agreement
is good. The major reason for the differences is the imperfect
modelling of the real start-up characteristics of the compressor.

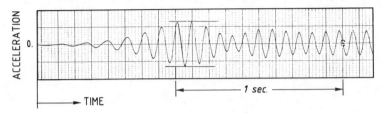

(a) Measured vertical acceleration during start-up.

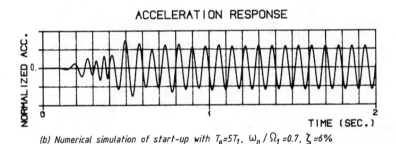

(b) Numerical simulation of start-up with $T_0 = 5T_1$, $\omega_n / \Omega_1 = 0.7$, $\zeta = 6\%$

Fig. 11   Qualitative comparison of measured response and numerical
          simulation

7. CONCLUSIONS

Low-tuned machine foundations for two compressors were designed for a
site with poor soil conditions. The special design of the foundation
piles allowed for accurate estimation of the foundation stiffness and
very low vibration levels at nearby facilities. The design, construction, testing and analyses of the foundations led to the following
conclusions:

1. Low-tuned foundations provide economical solutions for machine
    foundations in certain situations.

2. Vibration measurements are essential in checking the foundation
    performance, identifying the inconsistencies between design and
    construction, and identifying and correcting the inadequacies of
    the design.

3. Passing through resonance during start-up and shut-down did not
    lead to large transient vibrations for these foundations.

Acknowledgement

The foundations were designed for Norsk Hydro A/S. Their enthusiastic
participation in design and testing of this type of foundations, and
their willingness to release the present information for publication is
greatly appreciated.

A P P E N D I X   I

REFERENCES

1. Bathe, K.-J. (1982)
    Finite element procedures in engineering ananlysis. Englewood
    Cliffs, N.J., Prentice-Hall. 735 p.

2. Crede, C.E. (1965)
    Shock and vibration concepts for Standardization (1974)
    Mechanical vibration of machines with operating speeds from 10 to
    200 rev/s - Basis for specifying evaluation standards. 9 p. ISO
    Standard 2372.

3. Hvorslev, M.J. (1949)
    Subsurface exploration and sampling of soils for civil engineering
    purposes. Waterways Experiment Station, Vicksburg, Miss. 521 p.

4. Verein Deutscher Ingenieure (1964)
    Beurteilungsmasstäbe für mechanische Swingungen von Machinen.
    VDI-Richtlinien 2056.

A P P E N D I X   II

NOTATIONS

d	=	horizontal distance from C.G.
D	=	horizontal pile spacing
e	=	moment arm of eccentric mass
f	=	frequency
F_e	=	unbalanced force
F_t	=	dynamic force transmitted through the piles
I	=	moment of inertia of foundation plus machinery
k	=	axial stiffness of one pile
M	=	mass of foundation plus machinery
M_e	=	unbalanced moment
m_e	=	eccentric mass
p_o'	=	in situ vertical effective stress in soil
RMS	=	root-mean-square
s_u	=	undrained shear strength of soil
t	=	time
T	=	natural period
T_o	=	time required to reach steady-state operating frequency
$x, \dot{x}, \ddot{x}$	=	vertical displacement, velocity, acceleration
$\omega_n, \omega_{n_r}, \omega_{n_t}$	=	natural angular frequency (in rocking and translation)
Ω, Ω_1	=	excitation frequency, operating frequency
ζ, ζ_r, ζ_t	=	critical damping ratio (in rocking and translation)

Vibration Isolation of Machine Foundations

Dimitri E.Beskos* M.ASCE
Biswajit Dasgupta**
Ioannis G.Vardoulakis**M.ASCE

The problem of isolating a machine foundation by trenches, which reduce in amplitude the surface waves generated by the motion of the foundation, is numerically studied by the Boundary Element Method (BEM) in the frequency domain under conditions of plane strain. The soil medium is assumed to be a homogeneous isotropic and linear elastic half-plane,while the footing is assumed to be rigid, massive or massless,surface or embedded and subjected to harmonic forces. The trench is rectangular and may be either open or filled with some other materials, such as concrete. Both active and passive isolation cases are considered. Some parametric studies are also performed in this work to assess the importance of the various geometrical, material and dynamic input parameters and assist the engineer in successfully designing his isolation system.

Introduction

During the operation of large machines considerable dynamic forces, usually of a harmonic time viriation,are exerted on their foundations which in turn transmit harmonic waves into the supporting ground. These waves, which are primarily Rayleigh waves near the ground surface, radiate away from the foundation-source and adversely affect nearby structures and people. Structural protection is accomplished by various types of wave barriers, such as trenches, which are placed either close to or surrounding the source of vibration (active isolation) or far away from the foundation but close to the structure to be protected (passive isolation).The position of the wave barriers, their dimensions and possibly their material properties have to be selected in such a way so that the vibration isolation system to be effective, i.e., capable of considerably reducing the amplitude of the propagating waves. This requires the solution of a wave diffraction problem which cannot be easily solved analytically because of its complexities and resort should be made to either

* Professor, Department of Civil Engineering, University of Patras, Patras,Greece.
** Graduate Student and Associate Professor, respectively; Department of Civil and Mineral Engineering, University of Minnesota, Minneapolis, MN 55455.

experimental techniques or numerical methods of solution.
A comprehensive account of analytical solutions of simple
elastic wave diffraction problems can be found in the book
of Pao and Mow (13). The most comprehensive experimental
work on the suject of vibration isolation by trenches has
been done in the field by Woods(20)and a thorough exposition of this work together with all the pertinent references on the subject up to 1969 can also be found in the
classical text of Richart, Hall and Woods(14). Experimental methods of solution for this kind of problems are in-
valuable as they provide validation to the theory and very useful qualitative results, but are very expensive and
difficult to perform. What is needed is a numerical method
of solution capable of solving this problem accurately and
efficiently. Numerical methods can also help in conducting
extensive parametric studies easily and inexpensively. Waas(18),Haupt
(7,8),Segol et al(16) and May and Bolt(12)developed special
Finite Element Methods(FEM's) to study under conditions of
plane strain the problem of the amplitude reduction of surface waves by open or infilled trenches. Fuyuki and Mat-
sumoto(6) also studied the same problem by employing a special
Finite Difference Method (FDM). In the above references
the soil medium was assumed to be an isotropic,linear elastic
or viscoelastic, homogeneous or layered half-plane.However,
both the FDM and the FEM present a basic disadvantage in
dealing with wave propagation problems in soils in that
they represent a semi-infinite medium by a finite size model. Remedies to this may be either a very large, and hence uneconomical mesh, or use of complicated nonreflecting
boundaries(14,18,7,8,16,12) which are sometimes applicable
only when the layered soil is supported on a rigid bedrock
base(14,8). In addition, these methods badly simulate dynamic stress concentrations in problems of wave diffraction
as it has been demonstrated by Manolis and Beskos(10,11).

The Boundary Element Method (BEM) which is employed
in this paper for the study of the effectiveness of open
or infilled trenches as wave barriers, is free of these
disadvantages as requiring only a small portion of the soil
surface to be discretized and as taking automatically into
account the radiation condition. Furthermore, it is a
highly accurate and very efficient method, especially for
vibration isolation problems where one is primarily interested in soil surface displacements. Various special
forms of the BEM in the frequency domain have been used
for the solution of simple problems involving elastic wave diffraction by soil surface or subsurface topography
(5,2,19,15). Use of the direct BEM has also been employed
for the solution of wave scattering by two-and three- di-
mensional rigid embedded foundations in the frequency domain by Dominquez (4) and in the time domain by Spyrakos
and Beskos (17) and Karabalis and Beskos(9).

In this paper the problem of passive or active isolation of surface waves generating machine foundations by

open or infilled trenches of arbitrary shape is numerically
studied by the direct BEM in the frequency domain. The soil
medium is assumed to be a linear elastic or viscoelastic,ho-
mogeneous and isotropic half-plane. The developed methodology is capable of treating the "foundation-free soil sur-
face-trench" dynamic system as a whole. This methodology
utilizes an improved version of the type of the BEM described by Manolis and Beskos(10). For reasons of simplicity
constant boundary elements and the infinite plane Green's
function are employed in this work. This Green's function
requires, of course, a discretization of some portion of the
free surface of the soil medium in addition to the soil-foundation interface and the perimeter of the trench. Some parametric studies are also performed in this work to assess
the importance of the various parameters of the problem and
develop useful guidelines for the designer.

Frequency Domain Elastodynamics by the BEM

        Consider a homogeneous, isotropic and linear elastic
body B with boundary S under conditions of plane strain.Its
equation of motion in the frequency domain under the assumption of zero body forces takes the form

$$(c_1^2 - c_2^2) u_{i,ij} + c_2^2 u_{j,ii} + \omega^2 u_j = 0 \tag{1}$$

where u_i are the amplitudes of the components of the displacement vector, indices i and j correspond to Cartesian coordinates x and y, commas indicate differentiation with respect to these coordinates, summation convention is adopted
over repeated indices, c_1 and c_2 are the compressional and
shear wave propagation velocities, respectively, given in
terms of the Lamé constants λ and μ and the mass density γ
of the body by

$$c_1^2 = (\lambda + 2\mu)/\gamma, \quad c_2^2 = \mu/\gamma \tag{2}$$

and ω represents the circular frequency of the harmonically
varying with time motion of the body. The amplitudes of the
stresses σ_{ij} are connected ot those of the displacements u_i
by

$$\sigma_{ij} = \gamma \left[(c_1^2 - 2c_2^2) u_{k,k} \delta_{ij} + c_2^2 (u_{i,j} + u_{j,i}) \right] \tag{3}$$

where δ_{ij} is the Kronecker's delta. In vibration isolation
problems one has mixed type boundary conditions of the form

$$\sigma_{ij} n_j = \bar{t}_i \quad \text{on } S_\sigma, \tag{4}$$

$$u_i = \bar{u}_i \quad \text{on } S_u, \tag{5}$$

where n_j stands for the components of the outward unit nor-
mal vector at the boundary $S = S_\sigma + S_u$ and $\bar{t}_i$ and $\bar{u}_i$ represent

prescribed boundary values for the traction and displacement vectors, respectively. It is very easy to   observe
that the above frequency domain formulation of elastodynamics could have been obtained from its Laplace   transform domain formulation (10) by simply replacing the Laplace transform parameter s by iω, where i=√-1.

In order to solve the system of Eqs. 1 and 3-5   by
the BEM,use is made of the boundary integral equation(10)

$$\frac{1}{2} u_j(P) = -\int_S u_i(Q) \, T_{ji}(Q,P) \, dS(Q) + \int_S t_i(Q) U_{ji}(Q,P) dS(Q) \quad (6)$$

which is valid for points P and Q lying on a smooth boundary curve S and in which U_{ij} and T_{ij} are the   singular
influence tensors (fundamental solution or Green's   functions) for the infinite plane given explicitly, for example, in (10) with s being replaced by iω. These   tensors
are functions of ω and r, the distance between   the   two
boundary points P and Q and as r→0 become

$$U_{ij}=0(\ln r), \qquad T_{ij}=0(1/r) \qquad\qquad (7)$$

demonstrating their singular character.
The solution of Eq. 6 is accomplished numerically.   For
this purpose the boundary S of the body B is   discretized
into a number of N, in general unequal, straight segments
or boundary elements over which the stresses and displacements are assumed to be constant for reasons of   simplicity. Thus, Eq.6 can reduce in its discretized   form
to the matrix equation

$$T \, u = U \, t \qquad\qquad (8)$$

where u and t are the displacement and traction   vectors,
respectively, and T and U are influence matrices   consisting of entries which are integrals of the U_{ij} and T_{ij}
tensors, respectively, over the various constant boundary
elements.

In connection with the aboce BEM formulation   the
following two observations should be made:a) The vibration
isolation problems treated in this paper deal exclusively
with boundary quantities and thus there is no need for the
computation of internal displacements and stresses and b)
Use of the infinite plane Green's functions in the present
vibration isolation problems requires a discretization of
not only the soil-foundation interface and the   perimeter
of the trench but a small portion of the free soil   surface as well. Use of Green's functions defined for   the
half-plane do not, of course, require any free soil   surface discretization but they are very complicated (5).

Various details concerning the computational aspects

of the above formulation, such as element integration with
regular (P≠Q) and singular (P=Q) integrands, treatment   of
corners, special algorithms for solving a system of fully
populated linear equations with complex arithmetic, etc can
be found elsewhere (1,3).

Vibration Isolation of Footings by Trenches

          This section describes the formulation and solution
of the problem of passive and active vibration   isolation
of rigid massive machine foundations subjected to harmonic
forces with the aid of the frequency domain BEM. The amplitude reduction of the waves generated by the motion of the
foundation is accomplished by open or infilled trenches,
usually of rectangular shape, eventhough any other kind of
shape can be also treated by the proposed method. Figs. 1
and 2 show typical cases of passive and active   vibration

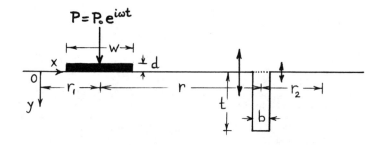

Figure 1.Passive vibration isolation of machine foundation
         by open or infilled trench.

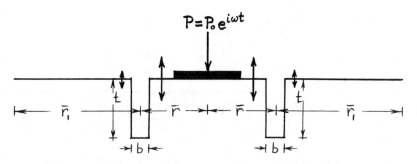

Figure   2.Active vibration isolation of machine foundation
          by open trenches.

isolation of machine footings by trenches, respectively. It
has been proven (14) that Rayleigh waves are the most important ones in problems of vibration isolation of foundations.

The present method, however, unlike some of the previous
ones, takes into account all the waves generated by the
motion of the footing and not just the Rayleigh waves,
exactly because it treats the foundation-trench system
as a whole.

Consider the passive vibration isolation problem
schematically described in Fig.1 and associated with the
open trench case. The footing, which is assumed to be
rigid, massive and bonded on the soil surface, is subjected to a vertical force P of the form

$$P = P_o e^{i\omega t} \tag{9}$$

The whole dynamic system of Fig.1 consists of the soil
half-plane with the trench and the rigid footing. A portion of the free soil surface and the trench perimeter
are discretized into a number of constant boundary elements, while the rigid footing is also divided into as
many elements as the corresponding soil ones at the soil-
foundation interface. The discretized portion of the
free soil furface, as Fig.1 indicates, extends over the
length $(r_1 + r + r_2) - (w + b)$. The matrix Eq.8 for the soil
surface takes the form

$$\begin{bmatrix} T_{ff} & T_{fr} \\ T_{rf} & T_{rr} \end{bmatrix} \begin{Bmatrix} u_f \\ u_r \end{Bmatrix} = \begin{bmatrix} U_{ff} & U_{fr} \\ U_{rf} & U_{rr} \end{bmatrix} \begin{Bmatrix} t_f \\ t_r \end{Bmatrix} \tag{10}$$

where the subscript f refers to quantities of the free
soil surface including the trench perimeter, while the
subscript r refer to those at the soil-foundation interface. For the free soil surface which is free of tractions one can write

$$t_f = 0 \tag{11}$$

For a rigid footing in vertical motion with a displacement amplitude Δ_y, the compatibility and equilibrium equations, in reference to the Cartesian coordinate system
shown in Fig.1 read

$$u_r = \begin{Bmatrix} u_{rx} \\ u_{ry} \end{Bmatrix} = \begin{Bmatrix} 0 \\ I\Delta_y \end{Bmatrix} \tag{12}$$

$$P_o = -m\omega^2 \Delta_y + \sum_{k=1}^{m} l_k \sigma_{yy}^k \tag{13}$$

where m is the mass of the foundation, l_k represents the
length of the k-th foundation element, $\sigma_{yy}^k = t_y^k$ are the interface tractions of the k-th foundation element and I
stands for the unit vector. Eqs.12 and 13 can be written
in a compact matrix form as

$$\underline{u}_r = \underline{\Delta} \tag{14}$$

$$\underline{F} = -\omega^2 \underline{M} \, \underline{\Delta} + \underline{L} \, \underline{t}_r \tag{15}$$

The system of matrix Eqs.10,11,14 and 15 can be easily solved and provide among others the vector $\underline{u}_f$. Knowledge of
the vector $\underline{u}_f$ permits one to determine the change in displacement amplitude for soil surface points before and after
the trench.

It should be noticed that the proposed   methodology is capable of treating not just the case considered above, which is the most usual one, but more involved   cases
including nonrelaxed boundary conditions, foundation   flexibility and embedment as well as rocking   and   horizontal
modes of footing motion in addition to the   vertical   one
considered here.

The problem of the active vibration isolation   of   a
machine foundation by open trenches as described schematically by Fig.2 is mathematically the same as   the   passive
one and can be analyzed in exactly the same way.

The case of the infilled trench, for both the passive and the active vibration isolation problems can be treated by simply supplementing the equations for the open trench
case with one additional boundary element equation for the
infill material in conjunction with the appropriate   equilibrium and compatibility equations at the soil-infill interface and the traction-free boundary condition   at   the
free top surface of the infill material. Thus, the system
of equations governing the problem consists of
a) the boundary equation for the soil medium

$$\begin{bmatrix} \underline{T}_{11} & \underline{T}_{12} & \underline{T}_{13} \\ \underline{T}_{21} & \underline{T}_{22} & \underline{T}_{23} \\ \underline{T}_{31} & \underline{T}_{32} & \underline{T}_{33} \end{bmatrix} \begin{Bmatrix} \underline{u}_f \\ \underline{u}_c \\ \underline{u}_r \end{Bmatrix} = \begin{bmatrix} \underline{U}_{11} & \underline{U}_{12} & \underline{U}_{13} \\ \underline{U}_{21} & \underline{U}_{22} & \underline{U}_{23} \\ \underline{U}_{31} & \underline{U}_{32} & \underline{U}_{33} \end{bmatrix} \begin{Bmatrix} \underline{t}_f \\ \underline{t}_c \\ \underline{t}_r \end{Bmatrix} \tag{16}$$

where the subscripts, $f,\bar{c}$ and r correspond   to   boundary
quantities associated with the free field soil surface,the
perimeter of the trench and the soil-foundation interface,
respectively;
b) the boundary equation for the infill medium

$$\begin{bmatrix} \underline{T}_{tt} & \underline{T}_{tc} \\ \underline{T}_{ct} & \underline{T}_{cc} \end{bmatrix} \begin{Bmatrix} \underline{u}_t \\ \underline{u}_c \end{Bmatrix} = \begin{bmatrix} \underline{U}_{tt} & \underline{U}_{tc} \\ \underline{U}_{ct} & \underline{U}_{cc} \end{bmatrix} \begin{Bmatrix} \underline{t}_t \\ \underline{t}_c \end{Bmatrix} \tag{17}$$

where the subscripts t and c correspond to boundary   quantities associated with the free top surface and the trench

perimeter-infill interface, respectively;

c) the soil free field boundary conditions

$$\underset{\sim}{t}_f^- = 0 \tag{18}$$

d) the compatibility and equilibrium equations at the soil-foundation interface, i.e., Eqs. 14 and 15;

e) the compatibility and equilibrium equations at the trench perimeter-infill interface

$$\underset{\sim}{u}_c^- = \underset{\sim}{u}_c \tag{19}$$

$$\underset{\sim}{t}_c^- = -\underset{\sim}{t}_c \tag{20}$$

f) the boundary condition at the free top surface of the infill material

$$\underset{\sim}{t}_t = 0 \tag{21}$$

The system of Eqs. 16-18,14,15 and 19-21 can be easily solved. Solution of this system provides the displacement vector $\underset{\sim}{u}_f^-$ which is needed for assessing the vibration isolation effectiveness of the infilled trench.

Numerical Examples

This section presents the numerical results obtained by applying the proposed methodology to solve two typical vibration isolation problems schematically described by Figs. 1 and 2.

a) Consider the passive vibration isolation problem of Fig. 1 in conjunction with the following numerical data: Material properties of soil medium: G_s=132 MN/m^2, v_s=0.25, $\bar{\gamma}_s$ =17.5 KN/m^2 c_{Rs}= 250 m/sec and ζ_s=6%; Material properties of infill (concrete): G_c =34.29 G_s, v_c=v_s, $\bar{\gamma}_c$=1.37 $\bar{\gamma}_s$, C_{Rc}=5 C_{Rs} and ζ_c=5 ζ_s, where $G,v,\bar{\gamma},C_R$ and ζ stand for shear modulus, Poisson's ratio, specific weight, Rayleigh wave velocity and hysteretic damping coefficient, respectively; Magnitude of applied force P_o=1 KN/m^2 and its operational frequency ω=50 Hz.

Figs. 3 and 4 depict the normalized vertical amplitude A(ξ) at the surface of the soil as a function of the dimensionless distance ξ=x/L_R, where L_R is the Rayleigh wavelength for the open and the concrete infilled trench cases, respectibely. A(ξ) is defined as the ratio of the amplitude when the trench is present to the amplitude in the absence of the trench. In both cases the normalized dimensions of the trench are τ=t/L_R=1.0 and β =b/L_R =0.1, the normalized dimensions of the foundation are π=w/L_R=0.5 and, δ=d/L_R=0.1, while the normalized distance between foundation and trench is ρ=r/L_R=5.0. A screening effectiveness measure is the amplitude reduction factor A_R defined as the

146 VIBRATION PROBLEMS

average normalized vertical surface amplitude behind the
trench over the length $r_2'=r_2-(b/2)$. In both Figs. 3 and

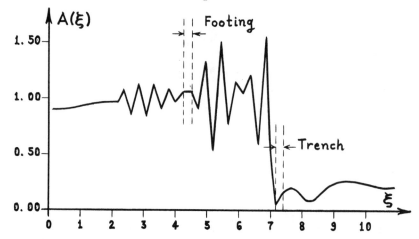

Figure 3. Normalized amplitude $A=A(\xi)$ for passive vibration isolation of footing by open trench.

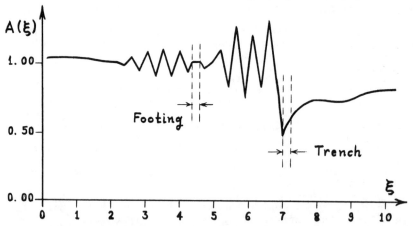

Figure 4. Normalized amplitude $A=A(\xi)$ for passive vibration isolation of footing by concrete filled
trench.

and 4 there is an amplitude reduction, but while $A_R\approx0.25$
for the open trench, $A_R\approx0.75$ for the infilled trench for
surface points after the trench. This clearly indicates
that, for the parameters chosen, the open trench performs
much better than the concrete filled trench. More specifically, according to the criterion of effectiveness

established in (14), the open trench in this example is considered to be a successful wave barrier becasue $A_R \leqslant 0.25$, while the filled one clearly represents a rather poor design.

(b) Consider the active vibration isolation problem of Fig.2 in conjunction with the material and dynamic input data of example (a). With geometrical parameters τ, β, π and δ those of example (a) and with $\bar{\rho} = \bar{r}/L_R = 1.0$ and $\bar{\rho}_1 = \bar{r}_1/L_R = 5.0$, the proposed methodology was utilized to construct Fig.5, which provides $A = A(\xi)$ for the case of open trenches and clearly indicates that the selected parameters resulted in a successful vibration isolation design for which approximately $A_R \leqslant 0.25$.

Parametric studies

In an effort to assess the screening effectiveness of both open and filled kinds of trenches in a general way detailed parametric studies were conducted on the basis of example (a). At the beginning a preliminary study revealed that the foundation mass does not practically affect A_R and, as a result of that, the assumption of massless foundation was adopted in the parametric studies (3). These studies were conducted by keeping constant the material parameters and the parameters ρ_1, ρ_2 and π and varying the parameters, τ, β, ρ and ω. The main conclusions of these studies, which are presented in detail in (3), are the following:
1) For open trenches A_R decreases for increasing τ, while it shows almost no dependence on β. The screening is successfull, i.e., $A_R \leqslant 0.25$ for $t \geqslant 0.6$. These conclusions are in agreement with (16).
2) For open trenches A_R is almost insensitive to ρ for all β, in agreement with (16).
3) For open trenches A_R decreases with increasing ω, i.e., with decreasing L_R in agreement with (16).
4) For infilled trenches A_R decreases for increasing β and τ. The dependence of A_R on β is very strong while on τ is very weak. For concrete filled trenches $A_R \leqslant 0.25$ for $\beta \cdot \tau \geqslant 1.50$. These results are in agreement with (7,8).
5) For infilled trenches A_R is almost insensitive to ρ.
6) In general, open trenches are more effective wave barriers than infilled trenches. Open trenches, however, present problems of instability of their walls and from this viewpoint concrete filled trenches are better.

Thus, by considering both the open and infilled trench cases, the above study was able to prove that the findings in (20,14,7,8,16) were completely consistent at least qualitatively.

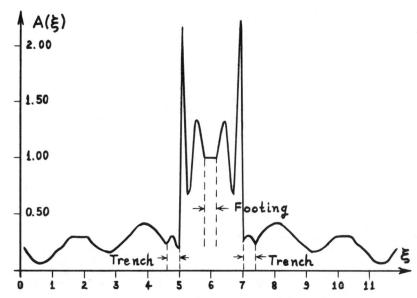

Figure 5.   Normalized amplitude A=A(ξ) for active vibra-
             tion isolation of footing by two open trenches.

Conclusions

        The main conclusions of this work are the follo-
wing:
1) A numerical methodology based on the frequency domain BEM has been developed for the very efficient
   treatment of vibration isolation of machine foundation problems under conditions of plane strain.
2) Both active and passive vibration isolation by open
   or infilled trenches have been considered and numerical examples pertaining to those cases have been
   solved to illustrate the proposed methodology.
3) Parametric studies have been conducted to assess the
   importance of the various geometrical, material and
   dynamic input parameters on the trench effectiveness
   as a wave barrier and provide some design guidelines
   to the engineer. The screening is effective for
   $\tau \geqslant 0.6$ for open and $\beta \cdot \tau \geqslant 1.50$ for concrete filled trenches
   where $\tau = t/L_R$, $\beta = b/L_R$ with t and b being the depth
   and width of the trench, respectively and L_R being
   the Rayleigh wave length. In general, open trenches
   are more effective than infilled trenches but they
   present wall instability problems.
4) More work is needed in this area of vibration isolation of machine foundations in conjunction with more
   realistic modelling capable of including soil layering and anisotropy and problem three-dimensionality.

The BEM has the potential of effectively treating problems with these additional complexities.

Acknowledgement

    The authors are grateful to the National Science
Foundation for supporting this work under Grant NSF/CEE
81-09723.

Appendix I-References

1. Beskos,D.E., Dasgupta, B. and Vardoulakis, I.G.,"Vibra-
    tion Isolation Using Open or Filled Trenches. Part I:
    2-D Homogeneous Soil", Earthquake Engineering and Stru-
    ctural Dynamics, submitted.

2. Chu, L.L., Askar, A. and Cakmak, A.S., "An Approxima-
    te Method for Soil-Structure Interaction for SH-Waves.
    The Born Approximation", Earthquake Engineering   and
    Structural Dynamics, Vol.9,1981, pp.205-219.

3. Dasgupta, B., "Vibration Isolation of Structures   on
    Homogeneous Soil", thesis presented to the Universi-
    ty of Minnesota, in Minneapolis, Minn.,in 1985, in
    partial fulfillment of the requirements for the de-
    gree of Doctor of Philosophy.

4. Dominguez, J., "Response of Embedded Foundations to
    Travelling Waves", Report No.R78-24, Department of Ci-
    vil Engineering, Massachusetts Institute of Technolo-
    gy, Cambridge, Massachusetts, August, 1978.

5. Dravinski, M., "Scattering of Elastic Waves by an
    Alluvial Valley of Arbitary Shape", Report No. CE 80-
    06, Department of Civil Engineering, University of
    Southern California, Los Angeles, California,1980.

6. Fuyuki, M. and Matsumoto, Y., "Finite Difference Ana-
    lysis of Rayleigh Wave Scattering at a Trench", Bul-
    letin of the Seismological Society of America, Vol.
    70, 1980, pp.2051-2069.

7. Haupt, W.A., "Isolation of Vibrations by Concrete
    Core Walls",Proceedings of the 9th International Con-
    ference on Soil Mechanics and Foundation Engineering,
    held at Tokyo, Japan, Vol.2,1977, pp.251-256.

8. Haupt, W.A., "Surface Waves in Non-Homogeneous Half-
    Space", in "Dynamical Methods in Soil and Rock Mecha-
    nics", B.Prange, Editor, A.A.Balkema Publishers, Rot-
    terdam, 1978, pp.335-367.

9. Karabalis, D.L. and Beskos, D.E., "Dynamic Response
    of 3-D Rigid Embedded Foundations by Time   Domain
    Boundary Element Method", Computer Methods in Applied

Mechanics and Engineering,to appear。

10. Manolis, G.D。 and Beskos, D.E., "Dynamic Stress Concentration Studies by Boundary Integrals and Laplace Transform", International Journal for Numerical Methods in Engineering, Vol.17, 1981, pp.573-599。

11. Manolis, G。D。 and Beskos, D。E。, "Dynamic Response of Lined Tunnels by an Isoparametric Boundary Element Method", Computer Methods in Applied Mechanics and Engineering, Vol.36,1983,pp。291-307。

12. May, T.W. and Bolt, B.A。, "The Effectiveness of Trenches in Reducing Seismic Motion", Earthquake Engineering and Structural Dynamics, Vol. 10, 1982,pp。195-210。

13。 Pao, Y.H. and Mow, C。C., "Diffraction of Elastic Waves and Dynamic Stress Concentrations", Crane Russak Publishers, New York,1973。

14。 Richart, F.E., Jr。, Hall, J。R.,Jr and Woods,R。D., "Vibrations of Soils and Foundations", Prentice-Hall, Englewood Cliffs, New Jersey, 1970.

15。 Sanchez-Sesma, F.J。, "Diffraction of Elastic Waves by Three-Dimensional Surface Irregularities, Bulletin of the Seismological Society of America, Vol。73, 1983, pp.1621-1636.

16。 Segol, G., Lee, P.C。Y。 and Abel, J。F., Amplitude Reduction of Surface Waves by Trenches", Journal of the Engineering Mechanics Division,ASCE, Vol.104,No EM3, Proc。Paper 13848, June, 1978, pp.621-641.

17。 Spyrakos, C。C。 and Beskos,D.E。, "Dynamic Response of Rigid Strip Foundations by Time Domain Boundary Element Method", International Journal for Numerical Methods in Engineering, to appear。

18. Waas,G。,"Linear Two-Dimensional Analysis of Soil Dynamics Problems in Semi-Infinite Layered Media", thesis presented to the University of California, at Berkeley,Calif。,in 1972,in partial fulfillment of the requirements for the degree of Doctor of Philosophy.

19。 Wong.,H.L。,"Effect of Surface Topography on the Diffraction of P, SV and Rayleigh Waves", Bulletin of the Seismological Society of America, Vol.72, 1982,pp.1167-1183.

20. Woods, R.D.,"Screening of Surface Waves in Soils", Journal of the Soil Mechanics and Foundations Division, ASCE, Vol. 94,No.SM4, Proc.Paper 6031,July, 1968,pp.951-979.

SETTLEMENT FROM PILE DRIVING IN SANDS

Hugh S. Lacy,[1] M. ASCE and James P. Gould,[2] M. ASCE

Abstract

The paper includes a review of factors influencing settlements from pile driving vibrations, a listing of certain references and a description of case histories from Mueser Rutledge jobs, five of these with bearing piles and four involving trench sheeting. Soils susceptible to densification by vibration are narrowly-graded, single-sized clean sands with relative density less than about 50 to 55 percent. In these materials, settlement and structural damage can result when peak particle velocities are much less than the 2 in./sec which are taken as the ordinary safe limit for buildings. Factors that increase the total vibration energy input will increase settlements. These include the number, length and type of piles in the new foundation and the requisite driving resistance. The mere size of a job can convert a case with insignificant vibration effects to one causing damaging settlements.

Introduction

This paper concerns settlement from pile driving in essentially cohesionless sands and silts. Damage to structures caused by soil movements resulting from pile driving in such materials can be more significant than structural damage due to transmitted vibrations. Previous studies have identified the influence of energy input and of distance from the driven pile. However, there is little information in case histories on soil properties in combination with site geometry and driving energy. The soil mechanics aspect of the subject is the focus of this paper. Case histories are presented involving late Wisconsin deposits in the northeastern states.

Bearing piles are driven with impact hammers and sheeting is installed with vibratory hammers at thousands of sites. Settlement effects are observed in only a small percentage of these cases; but these often result in costly damage to structures. Soil characteristic is the key element in creating a potential for settlement. Site geometry is a factor whose influence is often unrecognized. This includes the difference between the founding level of the adjacent structure and the excavation level from which piles are driven, the load intensity on foundations of the adjacent structure and the depth to which new piles will be driven below adjacent footings or piles. The mode of lateral support provided when driving piles within an excavation also influences ground movements.

[1] Sr. Assoc., Mueser Rutledge Consulting Engineers, 708 Third Avenue, New York, NY 10017
[2] Partner, Mueser Rutledge Consulting Engineers, 708 Third Avenue, New York, NY 10017

Causes of Settlement

Pile driving close to existing structures can result in unacceptable settlement due to densification or lateral movement of supporting soil. Structures on friction piles can experience a temporary loss of frictional resistance when subject to driving vibrations (Lynch, 13). Basic parameters are the energy released in the pile driving and the character of the supporting soil. The final illustration of this paper, Figure No. 11, assembles information from case histories cited herein and Dowding (8) on the peak particle velocity versus distance from the energy source generated for a variety of driving hammers and pile types. This relationship of velocity, distance and energy input is the starting point for assessment of pile driving effects. However, a simple relationship between energy input, distance and settlement cannot be expected because a variety of other factors intervene, including the total energy expended in the entire course of a pile foundation installation, the vulnerability of adjacent foundations and the increase of shear stresses in the ground caused by the new excavation.

Material properties of interest appear to be those traditionally associated with liquefaction potential. The range of gradation of sands susceptible to seismic liquefaction (Bhandari, 3) is shown on Figure No. 10 with gradations of sands involved in the case histories. Laboratory studies (Silver, 20) have demonstrated that volume decrease can occur at low cyclic strain amplitude after many repetitions, as with pile driving, as well as under relatively few cycles at large strain, as in a seismic disturbance.

Several relationships between factors that influence damage to structures from pile driving have been described by previous studies (1, 4, 6, 9, 11, 15, 16, 19, 21). Peak particle velocity, generally accepted as the critical parameter, is related to displacement by:

$$v = 2\pi fd \qquad [1]$$

v = peak particle velocity in inches per second
f = soil vibration frequency in hertz
    (Hz = cycles per second)
d = displacement or amplitude in inches

A maximum peak particle velocity of 2 in./sec (51mm/sec) frequently is specified as a criterion to prevent structure damage. Four in./sec (102mm/sec) is the threshold of actual damage. However, densification of loose cohesionless soils by pile driving often is more damaging than vibrations transmitted to the structure itself. Some of the references agree that structures settle from ground densification when peak particle velocities measured either on the structure or on the ground are substantially below 2 in./sec (51mm/sec).

While there appears to be no accepted criterion to limit vibration-induced settlements, Dalmatov (7) found that settlements at two building sites were insignificant if transmitted accelerations were less than 1 to 5 percent of gravity. Szechy (21) states that "settlement can be expected if accelerations exceed 10 percent of gravity". Particle velocity is related to acceleration:

$$v = a/(2\pi f) \qquad\qquad [2]$$

a = acceleration in inches/sec/sec

Particle velocities corresponding to accelerations of 1, 5 and 10 percent of gravity are 0.02, 0.10 and 0.2 in./sec (0.5, 2.5 and 5.0 mm/sec), assuming a typical vibration frequency for sands of 30 Hz.

Crandell (4) and Luna (12), working with large numbers of cases of vibrations transmitted from blasting and pile driving, recommended the traditional limits on peak particle velocity based on damage to the structure rather than on soil strains or displacements.

The peak particle velocity has been found to be a function of the square root of the hammer energy in foot-pounds (E) divided by distance in feet from the pile tip (D):

$$v = K \left[(E)^{\frac{1}{2}}/D \right]^{n} \qquad\qquad [3]$$

K is a dimensionless coefficient dependent on pile impedence which is the product of the pile area, its density and its sonic velocity (Heckman, 9). For illustration, a typical K of 0.1 is assumed. Piles being driven at 58 ft (18m) from a structure, using a 30 kip-ft (40,680J) hammer, would produce a peak particle velocity of about 0.3 in./sec (7.6mm/sec). The value n is often assumed to be 1.0 for sands (Wiss, 23) although commonly used plots often use n = 1.5 (Wiss, 24 and 25).

Impact Hammers - These impart blows to the pile at a frequency of ½ to 2 hertz. Rayleigh waves, produced by impact driving have an initial frequency dependent on soil type and stratigraphy but independent of hammer frequency. The amplitude of the resulting vibrations dissipate rapidly with time between each blow and with increasing distance from the pile. Several curves of peak particle velocity estimated from Equation [3] for typical impact hammer and pile combinations are plotted as exponential straight lines in Figure No. 11, taking the value of n = 1.0. Velocities measured in the cases discussed herein are plotted as individual points, zones or short segments, depending on the particular conditions of measurement. These values agree reasonably well with those predicted by Equation [3] for various hammers.

Vibratory Hammers - Driving by a vibratory hammer forces the ground to oscillate at the vibrator's operating frequency. Vibratory hammers have an operating frequency generally between 22 and 28 Hz with at least one type that can vary between 7 and 27 Hz. This continuous vibration is typically close to, or above, the natural frequency of the surrounding soil deposit in which sheet piles are driven. A vibratory hammer operating at a frequency that is resonant with the ground can produce a large increase in amplitude of ground motion. Potential for soil densification and resultant settlement are magnified. When the hammer operating frequency exceeds that of the soil in which the pile is being installed, resonance will be experienced each time the hammer is started or stopped. It is, therefore, important to minimize interruptions in hammer operations.

Wiss (23) recommended that the transmitted peak particle velocity for
vibratory hammers should be only 1/2 to 1/3 that permitted for impact
hammers, because sustained operation of the vibratory hammer can set
up a resonant response in buildings.

Review of Case Histories from References

Table No. 1 summarizes pertinent cases from references reviewed.
These include instances of significant settlement and damage to
adjacent structures with measured peak particle velocity between 0.09
and 0.4 in./sec (2 to 10mm/sec). Heckman (9) describes installation
of trench sheeting by vibratory hammer which resulted in large
settlement and damage to buildings. On an adjacent section of this
project, damage was avoided by using H-piles on 6 ft (1.8m) centers
and lagging. The peak particle velocity measured during installation
of the H-piles at a distance of 30 ft (9.2m) was only 0.02 in./sec
(0.5 mm/sec). Relatively few data are published to permit a
quantitive analysis. Cases P, Q and J through O (Dowding, 8) have
been plotted in Figure No. 11 which summarizes peak particle velocity
measured at distances from various vibration sources. Measured values
of peak particle velocity agree fairly well with those predicted for
the pile hammer used.

Case Histories from Mueser Rutledge Files

Nine cases are described below. Information available is
summarized on Table No. 2. In all cases, vibration-related settlement
was observed and significant damage resulted. Projects are located in
the northeast in three general areas: New York City, lower
Connecticut River and Syracuse, New York. Soils involved are late
glacial outwash sands and silts. Gradations tend to group in a fairly
narrow band as shown in Figure No. 10. Preconsolidation varies from
site to site, depending on the load imposed by renewed advance of
waning glacial ice.

Case A - Foley Square, New York City - An early intractable
example (1960's) of vibratory driving occurred in Foley Square in
southern Manhattan. This is the location of Collect Pond where up to
several hundred feet (100±m) of late Wisconsin outwash fine sand and
varved silt overlie bouldery till in a deep bedrock trough. Bearing
piles, 14HP73, were driven for a high-rise structure at 3 ft (0.9m)
centers through 80 ft (24m) of sand and varved silt for 75 ton (670kN)
capacity in till. The problem was studied by Tschebotarioff (22) when
impact driving with a 26 kip-foot (35,300J) hammer began to cause
settlement of adjacent buildings on footings above the glacial sand.
Properties of the sand are given in Table No. 2 and Figure No. 10. A
geological section through the pile driving operations and adjacent
buildings is shown in Figure No. 1.

Concern for 1 in. (2.5cm) settlements observed from impact
driving led to underpinning of the front wall of the adjacent low
building with shallow jacked piles as reactions for active jacks to
keep the facade immovable. As settlements continued, a decision was
made to employ the recently available "sonic" and "sub-sonic"
vibratory hammers, assuming that less total energy and time would be

TABLE NO. 1

CASE HISTORIES FROM REFERENCES

CASE DESIGNATION (REFERENCE)	DRIVEN PILE OR SHEETING	SOURCE OF VIBRATION				SOIL DATA	REMARKS
		HAMMER	INPUT ENERGY (FT-KIPS)	DISTANCE PILE TO MEASUREMENT (FT)	PEAK PARTICLE VELOCITY (IN./SEC.)		
P (5)	PZ (Belgium)	ICE 812	4	5 TO 150	2 TO .03	Loose fill with boulers soft clay & dense sand	18Hz. Dr=35 to 65%. Plots of vertical strain versus ground accelaration for various values of Dr. Settlement up to 6".
U (7)	PZ (Russian)	Drop	6.5	3 TO 21	-	Medium sand	Measured peak particle acceler-ations. Settlement 0.2 inches. Soil vibrating at 18 to 56 Hz.
V (7)	24"Hollow concrete	Drop	20	-	-	Fill over sandy silt	Soil vibrating at 24 Hz. No measured settlement of water main.
W (12)	Bearing piles	Percussive bored pile	-	10	-	-	Pile installation method avoided settlement from use of impact hammer.
X (12)	Bearing piles	Impact hammer	-	Very close	-	-	Piles at varying distances were driven with different restrictions.
Y (13)	12"Pipe & 14"shell	Vulcan OR	30	3.5 To 28'	-	Sand fill; organic silt loose to medium dense sand (N=25);limestone, compact sand.	Previously driven 60 to 80 ft.piles settled up to 7". Telltales showed vibration caused downdrag loading the pile tip to 35 tons. Required all piles be driven within 30 ft. before placing pile cap.
R (9)	14"Pipe	Link-belt 520 diesel	30	-	0.1 TO 04	Rubble fill;10'-30' silty clay; 30'+ loose silty sand; stiff silty clay	Piles driven with mandrel. Peak particle velocity did not increase when driving resistance in lower stiff clay increased.
S (9)	H-Piles for trench	Diesel	-	-	-	Asphalt and fill/ compact silty loam	Measured peak particle velocity was higher for 12" H-pile than for 14" H-Pile.
Q (9)	Steel sheeting	Vibratory	-	110 12	0.09 0.2	Loose to medium dense sand.	1/4" settlement 35' from trench. 1/8" settlement 90' from trench.
T (9)	12" H-pile for trench	MKT 9B3	9	30	0.02	Loose to medium dense sands.	Reported previous large settlement and extensive damage to buildings at another site with similar soils and same hammer with steel sheeting. No settlement with H-pile.

TABLE NO. 2

CASE HISTORIES FOR THIS STUDY

MRCE CASE DESIG-NATION	LOCATION	PILE TYPE	SOURCE OF VIBRATION				N BLOWS /FT	PROPERTIES OF STRATUM CHIEFLY INVOLVED				REMARKS
			HAMMER	INPUT ENERGY (FT-KIPS)	DISTANCE PILE TO MEASUREMENT (FT)	PEAK PARTICLE VELOCITY (IN/SEC)		D60 MM.	D10 MM.	U(3)	$D_r(1)$ 10	
A	Foley Square NYC	14HP73	Impact "Subsonic" Bodine "Sonic"	26 — —	20 20 20	0.19 0.14 0.19	22-40 29	0.02 0.32	0.005 0.10	4 3	42-49 53-57	Buildings settled 3"
B	Lower Manhattan NYC	18" open-end pipe	Vulcan 010	32	—	—	20-40	0.35	0.10	3	40-60	1.5' settlement of street
C	West Brooklyn NYC	14HP73	Vulcan 08	26	5-30	0.1	8-25	0.12	0.03	4	30-50 40-60(2)	Structure settled 3" as 40 piles were driven.
D	South Brooklyn NYC	10.75 closed-end pipe	Vulcan 08	26	10-80	0.9-0.1	21-35	0.26	0.13	2	40	Structure settled 3" as 220 piles were driven
E	Lower Conn. River	12HP53	MKT 10B3	13-20	3.5 c-c	—	20	0.42	0.10	4	40	Ground between piles settled 2.75 ft.
F	West Brooklyn NYC	Hoesch 134	ICE 812	4.0	3	—	27	0.12	0.03	4	48(2) 40-60	Building settled 2.4"
G	N.Syracuse NY	PZ-27	ICE 416	2.2	10-25	—	1	Sandy silt/coarse to fine sand			25	Ramp settled 3" as sheeting removed
H	Syracuse NY	PZ-27	ICE 812	—	4' from sewer	—	7	Fine sandy silt/fine to coarse sand			30	Sewer settled 6" as sheeting removed
I	S. Queens NYC	Hoesch 134	ICE 812	4.0	4' from sewer	—	25	0.40	0.10	4	45	Sewer settled 3" as sheeting removed.

NOTES:

1. Relative density based on Bazaraa.
2. Relative density from actual measurement.
3. U=Uniformity coefficient =D60/D10.

required to advance the piles to till. Consultants advised that in the later driving, the excavated site should be backfilled in order to decrease the seepage forces and shear stresses beneath the site perimeter.

Figure No. 1 shows values of peak particle velocity measured during various driving activities. Ground between bearing piles settled 1 ft (0.3m) during the driving and, in addition, boiling appeared along the sides of the piles. "Sonic" and "sub-sonic" vibrators were then used to drive piles close to the adjacent buildings, but settlement continued. The new building's large plan area and total load, plus a conservative final driving criterion, required an enormous input of driving energy over a nine month period. The build-up of positive pore pressures undoubtedly propagated over a wide area. Final measured settlement of non-underpinned footings totalled 2½ to 3 in. (6.4 to 7.6cm). As a consequence of the differential settlement, it became necessary to demolish the adjacent structures. Settlements probably were aggravated by high spread bearing pressures, about 5 tons/ft² (480kPa) on the 16 story building.

Seismographs indicated a typical frequency of the glacial fine sand of 30 Hz. Interestingly, as driving continued and quick conditions developed, the resonant value was halved to about 15 Hz. The key to the problem is the slightly contractive nature of the late glacial sand which graded into varved silt, the classic New York City "bull's liver". H-piles at 3 ft (0.9m) centers displace two percent of the soil volume. The 1 ft (0.3m) settlement would create densification of another two percent of the sand gross volume, making 4% volume decrease. Assuming the original density deduced in Table No. 2, this increased relative density from about 45% to 50%. These glacial sediments were overridden by waning ice and have been used by Parsons (17) for spread bearing of buildings up to about 30 stories. While the deformability under static loads is low, densification under prolonged vibrations can be significant.

Case B - Lower Manhattan, New York City - This large excavation occupied most of a city block and extended about 50 ft (15.3m) below grade and 25 ft (7.6m) below groundwater, as illustrated on Figure No. 2. Subsoils are similar to Case A, being only several blocks away. The single-sized glacial sands become finer with depth, grading to a sandy silt, with bouldery glacial till overlying bedrock. Typical gradation of the dominant sand is shown on Figure No. 10. High building loads required closely spaced, 200 ton (1780kN), 18 in. (46 cm) diameter, open-end pipe to bedrock. The designers stipulated sheeting and bracing shown at the left of Figure No. 2. The specified procedure was to progressively excavate and install bracing to subgrade followed by bearing pile installation. The contractor proposed to install the perimeter sheeting, excavate to a 20 ft (6.1 m) depth, leaving a supporting berm around the site perimeter and install the piles with a follower from this level. Using this procedure, it was found that the sheeting gradually moved inward and along one street displaced 2 ft (0.6m). This caused the street to settle as much as 1.5 ft (0.46m), requiring most utilities be taken out of service and caused minor cracking in buildings across the street.

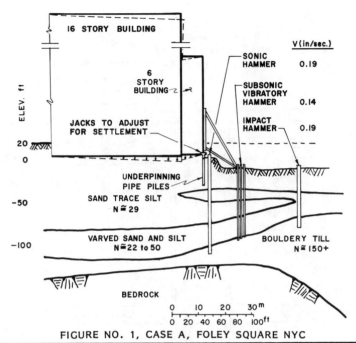

FIGURE NO. 1, CASE A, FOLEY SQUARE NYC

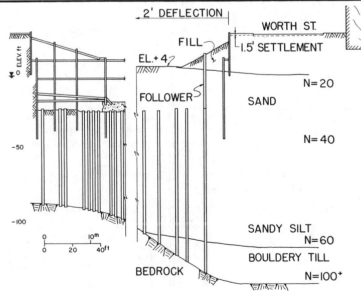

FIGURE NO. 2, CASE B, LOWER MANHATTAN, NYC

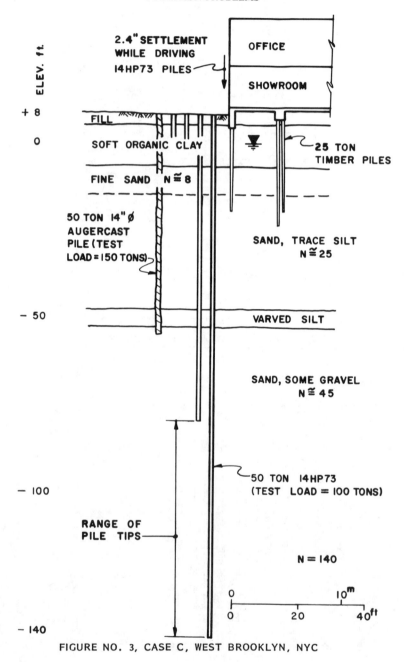

FIGURE NO. 3, CASE C, WEST BROOKLYN, NYC

A contributing cause may have been the difficulty in advancing pipe piles through the bouldery till above rock. Soil was repeatedly removed from inside the pipe and the pipe re-driven, probably causing loss of ground into the open pipe. However, it is believed that the primary cause of damage was vibrations during pile driving which increased pore pressures and reduced the berm's passive resistance to inward movement of the unbraced cofferdam.

Case C - Western Brooklyn, New York City - H-piles were driven at this site to construct a roadway ramp. The piles were 5 to 30 ft (1.5 to 9m) from an adjacent timber pile-supported building, as shown on Figure No. 3. Approximately 40 piles were installed before driving had to be terminated due to an accumulated settlement of 2.4 in. (6.1 cm) at one corner of the building. The new piles penetrated generally to depths of 100 to 150 ft (30 to 46m). The existing building timber piles had been driven to 25 ton (220kN) capacity, penetrating through organic soils, loose fine sand and a short distance into medium compact sand. Gradation of this bearing sand is shown on Figure No. 10. Maximum peak particle velocity measured on the building was 0.1 in./sec (2.5mm/sec). After consultation and load tests, 14 in. (36cm) diameter augercast piles were substituted for the remaining H-piles. No additional settlement of the adjacent structure was observed.

Case D - Southern Brooklyn, New York City - Expanding a treatment plant required construction of a structure next to existing aeration tanks and conduits, supported on 25 ton (220kN) timber piling. The base of these tanks is approximately 10 ft (3m) below surrounding grade. The length of the timber piles is unknown. Service loading was estimated to be 10 tons (89kN) per pile. The new structure extends 20 to 30 ft (6.1 to 9.2m) below grade, and necessitated lowering groundwater as much as 25 ft (7.6m) at elevator pits. A section through this site is shown on Figure No. 4. Properties of the sand are shown on Table No. 2, and the average gradation on Figure No. 10. Closed-end pipe piles at 47 tons (420kN) for the new structure drove to greater depths than anticipated before reaching the required driving resistance of more than twice that indicated by the Engineering News formula. Load test results, nevertheless, were slightly below that desired.

Elevations of adjacent structures were monitored from the beginning of construction. Significant settlement opposite where piles were being driven was first noted when about 100 new piles had been installed. Settlement continued at a relatively uniform rate to about 3 in. (7.6cm) until driving was halted when about 220 piles had been completed at distances ranging between 10 and 80 ft (3.1 and 24 m) from the edge of the aeration tanks. As shown on Figure No. 4, the settlement created little distortion in the tanks, being distributed over a 130 ft (40m) distance to the expansion joints at the center of the tanks. No vertical movement was experienced at the expansion joint and the opening at the top of the expansion joint was proportional to tank settlement opposite the pile driving. When settlement exceeded one inch (2.5cm), all fluid was removed from the tanks, reducing loading by about one half. However, the rate of

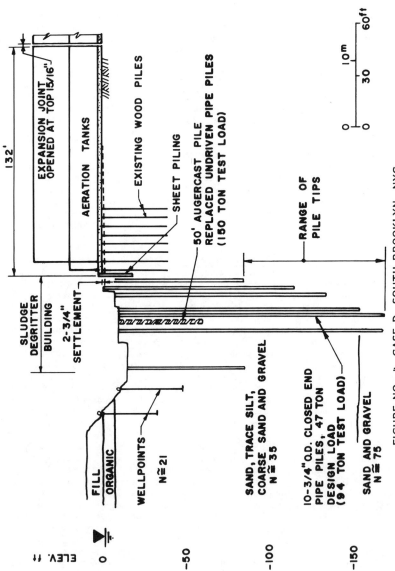

FIGURE NO. 4, CASE D, SOUTH BROOKLYN, NYC

settlement of the structure during the remaining pile driving was
unaffected.

During pile driving, a seismograph monitored vibrations on both
the structure and on the ground adjacent. Results are summarized on
Figure No. 11. Peak particle velocities initially measured on the
structure were consistently less than 0.1 in./sec (2.5mm/sec), while
those on the ground varied from 0.1 to 0.9 in./sec (2.5 to 23mm/sec).
Measurements of ground velocities generally agree with values
predicted for pipe piles with the Vulcan 08. There were several days
when two hammers were in operation. It was noted that when the
hammers became synchronous for several blows, the peak particle
velocity measured on the building suddenly increased by a factor of
about four.

The glacial outwash comprises chiefly single-sized, fine sands
with little passing the No. 200 sieve, as represented by the curve on
Figure No. 10. There was no indication of the generation of excess
pore pressure during driving as re-driving of piles at various levels
in the profile showed no significant change in driving resistance. It
is likely that operation of the dewatering system stifled possible
pore pressure build up.

Fourteen in. (36cm) diameter augercast piles were substituted for
the remaining 300 pipe piles at the same 47 ton (420kN) design load.
It was found that a 50 ft (15m) length successfully sustained 150 ton
(1340kN) test loads. Installation of the augercast piles produced no
significant reading on the seismograph. However, augering piles
immediately adjacent to the sheeting along the aeration tank resulted
in two additional inches (5cm) of tank settlement. The remaining 250
piles were installed without further incident. It appears that this
later settlement was not related to vibrations, but was caused by
reduction in passive resistance beneath the cantilever sheeting.

Case E - Lower Connecticut River - Construction for a major
highway bridge in 1945 to 1947 included conventional river pier
foundations where sheet pile cofferdams were installed within which
bearing piles were driven underwater, a tremie slab poured and the
pier built up. At a typical pier, 160, 12HP53 piles for 40 ton (360
kN) capacity were driven by an underwater 10B3 hammer to penetration
averaging 80 ft (24m) into a deep, remarkably homogeneous stratum of
single-sized sand filling the bedrock trough of the Connecticut River.
Gradation is shown on Figure No. 10 and driving data are given in
Table No. 2. A cross-section with profile of median spoon sampler
resistance and relative density (Bazaraa, 2) is shown in Figure No. 5.
Pile driving started in the center rows, working outward. As driving
progressed, pile lengths necessary to reach the requisite final
driving resistance increased. When soundings were made to determine
the base of the tremie pour, it was found that there had been an
average settlement of about 2-3/4 ft (0.84m) in the space between
piles. At 3.5 ft centers (1.1m), the piles occupy about 1.5% of the
enclosed soil volume. Settlement of 2-3/4 ft (0.84m) creates 3%
densification. The change in relative density for 4.5% compression
would be from about 40 to 47 percent.

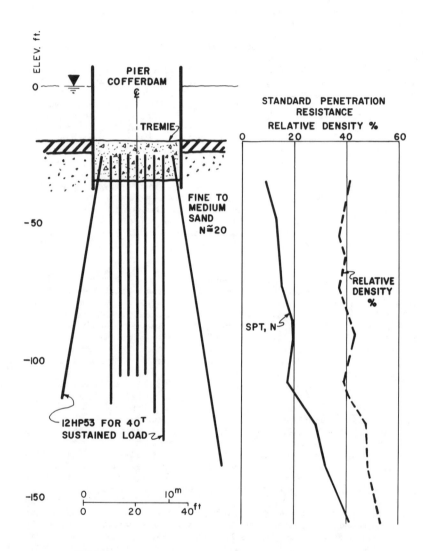

FIGURE NO. 5, CASE E, LOWER CONNECTICUT RIVER

While there was no transmission of this densification from pier to pier, it represents an extreme example of shake-down by vibration where small shear stresses were present. Because of this experience, the studies soon to be undertaken for a new parallel bridge emphasize the possible effect of new pile driving on the existing structure. Investigations will include measurement of ground movements and pore pressures produced by test pile driving. The experience is typical of cases in which deep sand compaction is achieved by the vibratory driving of open-end pipe piles, as in the "Terraprobe" procedure. Successful examples of this method have emphasized that the most appropriate material is a "saturated" uniformly graded SP sand lying in the band of those vulnerable materials shown on Figure No. 10.

Case F - Western Brooklyn, New York City - Steel sheeting was installed to permit an excavation next to a warehouse, supported on timber piling, as shown in the geological section on Figure No. 6. As the sheeting was being driven with a vibratory hammer, the wall of the warehouse less than 3 ft (0.9m) away settled progressively up to 3 in. (7.6cm), causing diagonal shear cracks in the grade beam between pile caps. Highly variable settlement in adjacent pile caps resulted in extensive cracking of the concrete block wall of the warehouse and leaning of the wall outward toward the new construction. Records of tip elevations for the piles supporting the warehouse indicate that the piles did not extend through the soft organic clay, but reached their required driving resistance in the fill. As the warehouse had no facility to resist lateral forces, the leaning concrete block wall supporting the roof truss was in danger of collapsing. This small warehouse was removed rather than being braced and underpinned and a new warehouse was constructed after the adjacent structure was completed.

Case G - North Syracuse, New York - Steel sheeting was installed for a bridge cofferdam. Subsequently, the adjacent area was filled, graded and pavement placed for an access ramp as shown on Figure No. 7. The sheeting was driven well below subgrade. Boils developed briefly in the excavation before a deep well was installed. The dewatering difficulties resulted from coarser permeable sands underlying very loose, fine sands and silts. During construction, the ramp settled significantly as a result of several contributing factors, including consolidation of silts below a filled former streambed. As a result, a joint in the adjacent sewer opened and surrounding soil was washed through the sewer causing large area settlement with the ramp settling up to 3 ft (0.9m) nearest the sewer. Careful records of elevations of the access ramp were kept during this period. Sheeting for the bridge cofferdam was pulled, resulting in rapid localized additional settlement of the ramp. Settlement was 3 in. (7.6cm) near the cofferdam and 1.5 in. (3.8cm) on the edge of ramp further from the sheeting.

Case H - Syracuse, New York - This project, illustrated on Figure No. 8, involved repair of an existing sewer where a joint had opened allowing silt and fine sand to enter. When the sewer became clogged, the damaged area was located by observing an 8 ft (2.4m) wide depression, 6 ft (1.8m) deep, for a 100 ft (30.5m) length, directly above the pipe. Steel sheeting 25 ft (7.6m) long was driven and the

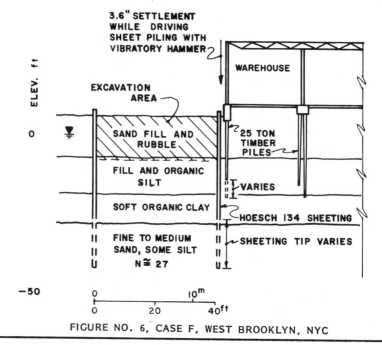

FIGURE NO. 6, CASE F, WEST BROOKLYN, NYC

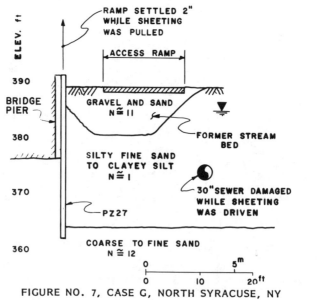

FIGURE NO. 7, CASE G, NORTH SYRACUSE, NY

damaged pipe was exposed. It was found to have settled as much as 30 in. (0.76m). Prior to excavation, deep wells were installed and the exposed subgrade appeared firm. Several feet of bedding was placed plus a concrete mat and cradle beneath the new pipe section. The trench was partially backfilled and sheeting extraction commenced only to discover that the sewer was settling a large amount. Sewer settlement reached 6 in. (15cm) after extracting sheeting along only 30 ft (9.2m) of the trench. Loosening of soil beneath the pipe invert, due to loss of soil into the open joint in the pipe prior to repair, probably contributed to the settlement of the newly installed sewer pipe. A cost analysis of leaving the sheeting in place indicated that a more economical solution would be to support the sewer pipe on timber piles.

Case I - Southern Queens, New York - Subsoils at this sewer construction consisted of the deep uniform, single-sized fine sand which is typical of the final glacial outwash on southern Long Island. Because of the fear of possible settlements which could be caused by systematic drawdown, the contractor was required to drive continuous steel sheeting a minimum of 18 ft (5.5m) below subgrade as shown on Figure No. 9; the intent being to reduce upward seepage gradients below critical levels. The Contractor was not permitted to dewater except from sumps. It was soon found that boils developed at subgrade, possibly as a result of the presence of coarser sands below the tips of sheeting. The crushed stone bedding below the pipe was thickened, and it appeared that the subgrade was stable. However, while pulling only nine sections of steel sheeting by vibratory extractor, the sewer settled up to 3 in. (7.6cm), and measurable settlement extended over a nearly 100 ft (30.5m) length of the sewer. A joint in the sewer pipe had opened an unacceptable amount. Apparently, a large part of the settlement was redensification of sand loosened by the upward seepage. Finally, a deep well dewatering system was installed and the sewer was excavated and re-installed. The trench sheeting already placed was not removed in this section because of concern that additional settlement might occur. The remaining sewer construction was accomplished by driving sheeting that extended 5 ft (1.5m) or less below subgrade and lowering groundwater below subgrade level with deep wells outside the sheeting. No significant settlement was experienced of the sewer constructed in dewatered ground although all sheeting was pulled. Measurements on adjacent structures confirmed that no measurable settlements accompanied groundwater drawdown.

Summary and Conclusions

This paper includes a brief review of factors influencing settlements from pile driving vibrations, a listing of certain references and a series of case histories from Mueser Rutledge job experiences, including five with bearing pile driving and four involving trench sheeting. Based on the case history information, the following conclusions are offered:

1. Settlement from pile driving in loose-to-medium compact narrowly graded sands can result from peak particle velocities much less than the 2 in./sec (51mm/sec) which is taken as the ordinary safe

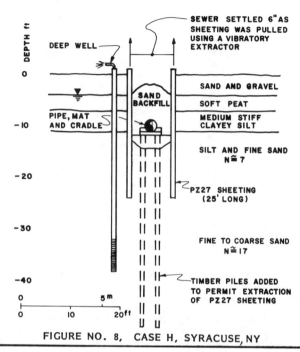

FIGURE NO. 8, CASE H, SYRACUSE, NY

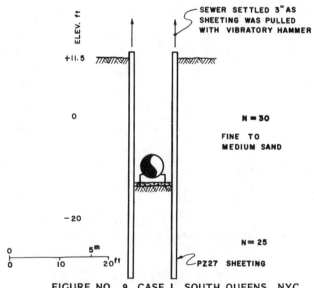

FIGURE NO. 9, CASE I, SOUTH QUEENS, NYC

limit for buildings. In fact, peak particle velocities as low as 0.1 to 0.2 in./sec (2.5 to 5.1mm/sec) measured on the ground surface appear to accompany significant settlements at some vulnerable sites. These values are of similar magnitude to those commonly measured from vehicular traffic on adjacent streets. However, vibrations from pile driving are propagated within the ground, directly into the vulnerable soil. The characteristics of susceptible sands are rather obvious and include late Pleistocene materials with uniformity coefficients ranging up to four or five and relative densities ranging up to 50 or 55 percent. Properties of lesser importance, whose influences are unclear are: effects of grain shape; soil permeability, as related to the build-up of excess pore pressures; anisotropy; magnitude of effective stress. Holubec (10) demonstrated that sands of angular grains experience a greater decrease of void ratio than those with rounded grain.

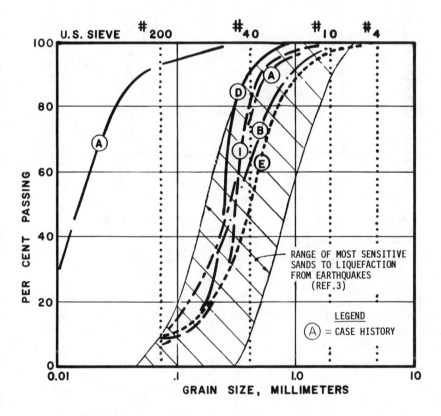

FIGURE NO. 10, GRADATION CURVES OF OUTWASH SANDS

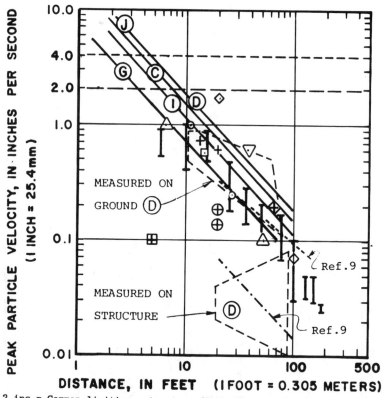

2 ips = Common limiting criterion

4 ips = Structural damage

ESTIMATED RELATIONSHIPS FOR CASES:

Ⓒ Vulcan 08, 26 KF, 12HP74

Ⓓ Vulcan 08, 26 KF, 10.75 Pipe

Ⓖ ICE 615, 2.3 KF, PZ 27

Ⓘ ICE 812, 4.0 KF, Hoesch 134

Ⓙ Vulcan 010, 32 KF, 14HP73

Pile impedence K= 0.1 for all
except Case D; K = 0.15

MEASURED VALUES FOR CASES:

⊕ Ⓐ 26 KF, 14HP73

⊞ Ⓒ

⬙ Ⓓ Measured on structure with
two hammers briefly
synchronized

⊡ Ⓙ

⊙ Ⓚ ICE 182, PZ 102

+ Ⓛ Foster 4000,43lb/ft.

△ Ⓜ LB 520D, 14WF89 ⎤
⎥
⎥ Ref.8
◇ Ⓝ Franki, 140 KF ⎥
⎥
▽ Ⓞ Drop wt, 170 KF ⎦

Ⅰ Ⓟ Ref. 5,ICE 812, PZ

---- Ⓠ Ref. 9,Vibratory, Driving
sheeting

—·— Ⓠ Ref. 9,Pulling Sheeting

FIGURE NO. 11, PEAK PARTICLE VELOCITY PRODUCED BY PILE DRIVING

2. Job characteristics that influence the settlement magnitude include: the distance between the source of vibration and the structure affected; a vulnerable position of adjacent foundations with respect to the new work; the position of the water table; and the presence of unbalanced hydrostatic head acting in and around the site. Since the settlement produced by pile driving is the result of the repetition of extremely small individual perturbations, those factors that increase the total vibration energy input, or the time span over which it is continued, will increase settlements. These include factors such as the depth of overburden, intensity of final driving resistance, number of piles and overall size of the site. In other words, the mere size of a job can change a situation from insignificant vibration effects to damaging settlements.

3. It is interesting to note that damaging particle velocities or acceleration are much lower than values associated with modest seismic events. Pile driving operations, which continue for many orders of magnitude longer than an earthquake, superpose very small effects for very many times to produce much greater settlements than those caused by the larger northeast earthquakes with accelerations of 0.05g to 0.10g. In fact, in vulnerable sands, the effect of pile driving is sometimes similar to that of a limited liquefaction in which materials are reported to have gone "quick". This is evidence that the subsoils are in a condition that can be densified by vibratory energy. However, it is not clear that the presence of excess pore pressures, which presumably decrease the natural frequency of a sand, positively increase the final magnitude of vibratory settlements.

4. Prediction of settlement produced in cohesionless sands is not now susceptible to a simple mathematical evaluation based on peak particle velocity. An informed judgement requires knowledge of gradation, relative density, site geometry, groundwater levels and hammer energy. The vulnerability of adjacent structures to settlements as opposed to vibrations transmitted to the structural frame itself must be judged. The scale of the project must be taken into account; that is, whether long-continued, hard driving is going to liquify a marginally contractive sand. In future site studies at least rudimentary in-situ density should be determined by proper "undisturbed" sampling employing orthodox Hvorslev procedures with a fixed piston sampler in a heavily mudded hole (Poulos, 18).

5. Traditional cautions in operating vibratory hammers are more or less confirmed by these case histories. Interruption in hammer operation is undesirable. In potentially damaging situations, it is beneficial to operate a vibratory hammer at a frequency removed from the resonant frequency of the sand. Where there may be serious consequences of settlement on vibration, pre-construction testing might include measurement of resonant frequency of the sands and a contract stipulation that these values are to be avoided by the operating frequency of the vibratory hammer.

6. The case histories of sewer construction indicate that it can be counter-productive to drive trench sheeting to great depths below invert in an effort to control inflow while avoiding exterior

drawdown. Sands which are susceptible to piping are also vulnerable
to settlement from vibration. Extraction of deeply penetrating sheets
can cause sewer pipe settlement, particularly if poorly controlled
seepage has loosened the subgrade soil. Alternatives might include
use of minimum penetration sheeting, driving with impact hammers,
using H-pile and lagging for trench support with dewatering procedures
which will absolutely prevent loosening of sands below invert.

7. A potentially dangerous condition arises in driving piles
through cohesionless materials of the type described above where that
material is relied upon for passive resistance against movements of a
cofferdam or excavated slope. Development of positive pore pressures
decreases the passive resistance or, in effect, decreases the spring
constant of the soil below subgrade. As a consequence, movements of
the retained surrounding ground can be drastically increased. Where
this is not a factor, it is not obvious that a non-displacement pile
has an advantage over a displacement pile in preventing settlement.
Displacement piles generate an increase in horizontal stresses which
would decrease the tendency for the surrounding ground to densify by
vertical compression. In fact, the bearing pile cases reported herein
include principally non-displacement piles. However, it is clear that
pre-coring or the use of uncased augered-in-place piles can reduce or
eliminate the threat of settlement due to driving vibrations.

8. Research is needed to establish the relationship between
cumulative energy input into a vulnerable sand stratum through driving
of numerous piles and the various other factors that contribute to
settlement described herein. Research in seismic liquefaction of
soils and in vibro-densification of soils should be useful in
developing an understanding of these relationships.

Appendix - References

1. Attewell, P.B., and Farmer, I.W., "Attenuation of Ground
   Vibrations from Pile Driving," Ground Engineering, Vol. 3, No. 7,
   July, 1973.
2. Bazaraa, A.R.S.S., "Use of the Standard Penetration Test for
   Estimating Settlements of Shallow Foundations on Sand," PhD
   Thesis for the University of Illinois, 381 pp., 1967.
3. Bhandari, R.K.M., "Dynamic Consolidation of Liquifiable Sands",
   Proceedings of the International Conference on Recent Advances in
   Geotechnical Earthquake Engineering and Soil Dynamics, St. Louis,
   MO, 1981, pp. 857-860.
4. Crandell, F.J., "Ground Vibration Due to Blasting and Its Effects
   Upon Structure," Contributions to Soil Mechanics, 1941-1953,
   Boston Society of Civil Engineers, Boston, MA, 1953.
5. Clough, G.W. and Chameau, J.L., "Measured Effects of Vibratory
   Sheet Pile Driving," Journal of Geotechnical Engineering
   Division, ASCE, Vol. 104, GT10, October, 1980, pp. 1081-1099.
6. Crockett, J.H.A., Hammond, R., "Reduction of Ground Vibration
   into Structures," Institution of Civil Engineers, Structures
   Paper No. 18, London, 1947.
7. Dalmatov, B.I., Ershov, V.A., and Kovalevsky, E.D., "Some Cases
   of Vibration Settlement in Driving Sheeting and Piles," Proc. of
   the Symposium on Wave Propagation and Dynamic Properties of Earth

Materials, University of New Mexico, Albuquerque, NM, 1968, pp. 607-613.

8. Dowding, C.H., Personal Communication, October 23, 1984.

9. Heckman, W.S. and Hagerty, D.J., "Vibrations Associated with Pile Driving," Journal of the Construction Division, ASCE, Vol. 104, No. CO4, Proc. Paper 14205.

10. Holubec, I. and D'Appolonia, E., "Effect of Particle Shape on the Engineering Properties of Granular Soils," Evaluation of Relative Density and Its Role in Geotechnical Projects Involving Cohesionless Soils, ASTM STP 523, American Society for Testing and Materials, 1973, pp. 304-318.

11. Lo, M.B., "Attenuation of Ground Vibration Induced by Pile Driving", Proc. of the 9th International Conference on Soil Mechanics and Foundation Engineering," Vol. 2, Paper No. 4/20, 1976.

12. Luna, W.A., "Ground Vibrations Due to Pile Driving," Foundation Facts, Vol. 3, No. 2, Raymond International, Houston, TX, 1967.

13. Lynch, T.J., "Pile Driving Experiences at Port Everglades," Journal of the Soil Mechanics and Foundation Division, Vol. 86, No. SM2, Proc. Paper 2442, April, 1960, pp. 41-62.

14. Mallard, D.J., Bastow, P., "Some Observations on the Vibrations Caused by Pile Driving," Recent Development in the Design and Construction of Piles, Institution of Civil Engineers, London, 1979.

15. Nielson, F.D., Cardenas, J.A., "Influence of Soil Properties on Volumetric Change under Dynamic Loading", Highway Research Record No. 181, Highway Research Board, Washington, DC, 1967.

16. O'Neill, D.B., "Vibration and Dynamic Settlement from Pile Driving", Proc. of the Conference "Behavior of Piles", The Institution of Civil Engineers, London, pp. 135-140, 1971.

17. Parsons, J.D., "New York's Glacial Lake Formation of Varved Silt and Clay," Journal of the Soil Mechanics and Foundation Division, ASCE, Vol. 102, No. GT6, June, 1976.

18. Poulos, S.J., Castro, G. and France, J.W., "Liquefaction Evaluation Procedure", Journal of Geotechnical Engineering Division, ASCE Vol. 111, No. 6, June, 1985, pp. 772-792.

19. Richart, F.E., Hall, J.R., and Woods, R.D., "Vibration of Soils and Foundations," Prentice-Hall, Inc., Englewood Cliffs, NJ, 1970.

20. Silver, M.L., Seed, H.B., "Volume Change in Sands During Cyclic Loading", Journal of Soil Mechanics and Foundation Division, ASCE, SM9, September 1971, pp. 1121-1182.

21. Szechy, K., Varga, L., "Foundation Engineering", Akademiai Klado, Budapest, 1978.

22. Tschebotarioff, G.P., "Foundations, Retaining and Earth Structures," 2nd Edition, McGraw-Hill Book Co., Inc., New York, NY, 1973.

23. Wiss, J.F., "Damage Effects of Pile Driving Vibration," Highway Research Record No. 155, Highway Research Board, Washington, DC, 1967.

24. Wiss, J.F., "Vibrations During Construction Operations," Journal of the Construction Division, ASCE, Vol. 100, No. CO3, Proc. Paper 10798, September, 1974, pp. 239-246.

25. Wiss, J.F., "Construction Vibrations, State-of-the-Art", Journal of the Geotechnical Engineering Division, ASCE, February, 1981.

PILE DRIVING INDUCED SETTLEMENTS OF A PIER FOUNDATION

by Miguel Picornell[1] M. ASCE and Evaristo del Monte[2]

ABSTRACT: Upon driving steel H-piles for the equipment foundations of a steel mill factory, one of the pier foundations of the building settled 10 in. (25 cm). The ensuing field investigation revealed a predominatly granular soil deposit, from loose to med. dense, with frequent huge limestone boulders. A large number of the building piers were found to rest on these boulders. The field tests show that, despite the low densities, the soil deposit is fairly incompressible and that the static design loads would cause only minimal settlements. These static settlements contrast with the settlements that occurred upon driving the H-piles, which are attributed to the dynamic compaction of the subsoil induced by the pile driving activities.

INTRODUCTION

The steel mill factory under investigation is located in Lesaka, northern Spain, in a very narrow valley enclosed by the steep hills of the west end of the Pyrennees. The expansion of this mill was intended to complement the existing facilities and was designed to handle much heavier products.

The new facilities cover an area of about 270,000 ft^2 (25000m^2) split into four workshops. These are made up of modules 39.6 ft (12 m) long and spans of 114.8 ft - 131.2 ft (35 m - 40 m). A sketch of the relative position of these areas is shown in Fig. 1 along with the location of the site investigation that will be described later.

The buildings have a steel structure. The steel columns are anchored into concrete footings. Each footing is supported on two cast-in-place concrete piers of 3.54 ft (1.08 m) in diameter, that are placed 9.7 ft (3 m) center-to-center.

The pier foundations had been designed to rest on bedrock. The design loads on the piers were 423 Kip-683 Kip (1.9 MN - 3.0 MN). These loads included 55 Kip - 110 Kip (245 kN - 490 kN) of negative skin friction. The equipment foundations had been designed on steel H-piles to be driven to bedrock. The maximum net load on the equipment mats was 3.1 Kip/ft^2 (147 kPa). The maximum expected surcharge in the stockpiling areas was 4.3 Kip/ft^2 (206 kPa).

---

[1]Assistant Professor, University of Texas at El Paso.

[2]Senior Eng., Tecnicas y Proyectos, S.A., Madrid, Spain.

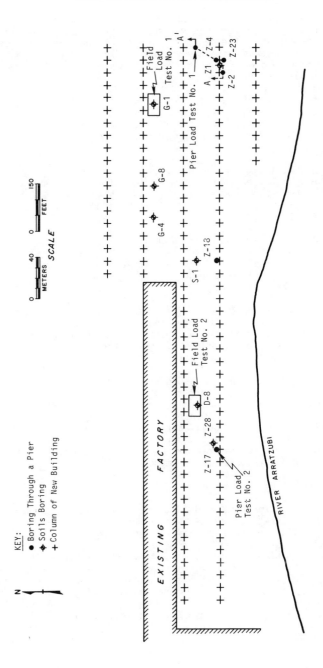

FIGURE 1, SKETCH OF FACTORY EXPANSION AND LOCATION OF RELEVANT SITE INVESTIGATION.

The expansion was initiated with the construction of the building. The first problem was encountered during the installation of the concrete piers. The depth of the excavation needed to reach the presumed bedrock was extremely variable. The difference in length of adjacent piers was more than 16 ft (5 m) in many cases and in a few instances exceeded 60 ft (20 m).

The second stage of the construction was the installation of the equipment. The first construction was at the eastern edge of the expansion near the footing that is shown in Fig. 1 with the borings Z4 and Z23. This footing was resting on two piers 62 ft and 69 ft (19 m and 21 m) long.

The mat for the equipment in the first construction was to rest on only a few H-piles, but these were located adjacent to the building footing. The length of H-pile that had to be driven to reach refusal was several times larger than the length of the concrete piers supporting the footing.

Before the driving was completed, it was noticed that the building footing had settled about 10 in. (25 cm). At that time, the only loads on the piers were the weights of the concrete and the structure. The static load on each pier could not have been more than about 110 Kip (490 kN).

The rest of the nearby footings remained apparently unaffected. A visual inspection, at the time when the movements were noticed, did not reveal the presence of cracks or any other feature on the ground surface that would suggest that the movements were the result of some type of cave-in effect.

These settlements prompted a reconsideration of the need to drive piles and to look for alternatives for the equipment foundations. At the same time, many uncertainities arose regarding the safety of the existing pier foundations. An additional site investigation was then implemented to address both aspects. A summary of these investigations is presented below, since they shed some light on the possible cause of the settlements and they help to show to what extent the pile driving was responsible for them.

SUBSURFACE CONDITIONS

The site investigation for the equipment mats included 38 borings drilled at the sites of the heavier slabs. The location of the investigation mentioned here is shown in Fig. 1.

The most striking feature of the site is the extremely irregular thickness of the soil deposit. While bedrock is outcropping in the slope that bounds the expansion on the North, elsewhere in the expansion bedrock was not reached at depths of 272 ft (83 m). Sizeable differences were even found between borings located less than 82 ft (25 m) apart. In boring Z18 bedrock was found at 33 ft (10 m) depth, but boring S1 was stopped at 230 ft (70 m) depth without reaching bedrock.

The soil deposit is predominately granular. In broad terms, the subsoil can be grouped into two main strata. The uppermost stratum consists of med. dense gravel with some sand and silt. The thickness of this stratum ranges from 3 ft to 33 ft (1 m - 10 m) over most of the area, except in the north-east corner of the expansion where it reaches more than 98 ft (30 m). Embedded in this stratum, there are occasional irregular lenses of med. stiff to stiff silty clay (liquid limit 30% and plasticity index 9%).

The deeper stratum consists of successive layers of silty sand and sandy silt with variable amounts of gravel and widely different densities. This stratum contains huge boulders or ledges of limestone up to about 20 ft (6 m) thick. In some cases, the soils underlying the boulders are extremely loose; for example, in a standard penetration test implemented at 100 ft (30 m) depth in boring G4, the sampler was driven about 7 ft (2 m) with three blows. The higher blow counts registered in this stratum usually correlate with the presence of gravel, so, it is thought that the overall density of this stratum is only from loose to med. dense.

As a general trend, the soil deposit becomes slightly less dense towards the eastern edge of the expansion. The ground water table in the area sloped gently in this direction from a depth of about 3.3 ft (1 m) on the western boundary to a depth of 16.4 ft (5 m) on the opposite end. A few schematic boring logs, that for clarity have been simplified, are presented in Fig. 2.

SOIL TESTS AND ANALYSIS

The compressibility under static loads of the two strata was investigated with laboratory tests and two field load tests. A representative sample of consolidation tests on specimens from this site is shown in Fig. 3 and the more relevant settlement records of the two field load tests are presented in Fig. 4. Load test no. 1 was used to assess the compressibility of the gravel layer and load test no. 2 was mainly intended to test the compressibility of the deeper layer of silty sand.

The field load tests were implemented by stacking up steel sheet coils of known individual weights. These were piled to an average stress of 3.1 Kip/ft^2 (147 kPa) over an area of about 66 ft x 49 ft (20 m x 15 m). The displacements were monitored with eight settlement plates per test. The plates were placed at 1.6 ft (0.5 m) depth near the center and the corners of the loaded area and a few were located at several distances outside the loaded area. In test no. 1, two of the plates near the center of the pile were placed at 20 ft (6 m) depth, just below the upper lens of CI-MI shown in boring G1.

The consolidation test on the CI-MI sample of boring G1 indicates that this layer has a preconsolidation pressure which is 4 to 8 times the existing overburden. In this test, the soil has a recompression ratio, C_{cr}, of 0.013. The recompression ratios backfigured from the settlements measured in the field load test no. 1, assuming that they are constant with depth, are as follows:

Center plate....................0.015
Corner plate...................0.0149
Center deep plate.............0.014

As it can be seen, there is a remarkable agreement between lab and field tests and both confirm the low compressibility of the gravel layer; furthermore, these results indicate that the compressibility of the Cl-Ml lense is of the same order of magnitude as the compressibility of the gravel layer under the expected loads of the mats and the stockpiling areas.

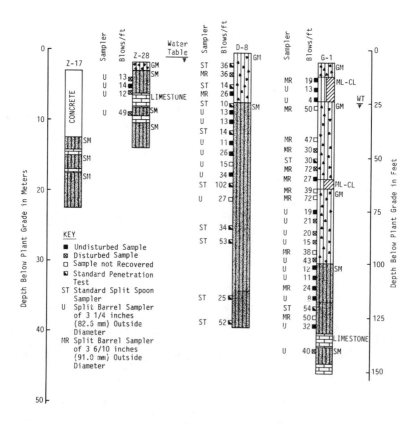

FIGURE 2, SUMMARY OF SOILS BORINGS LOCATED NEAR THE SITES OF OTHER FIELD TESTS.

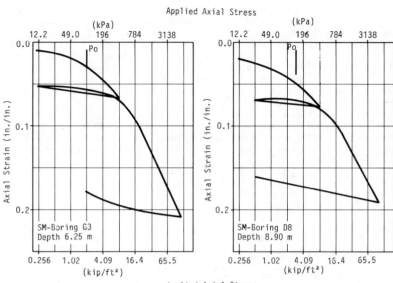

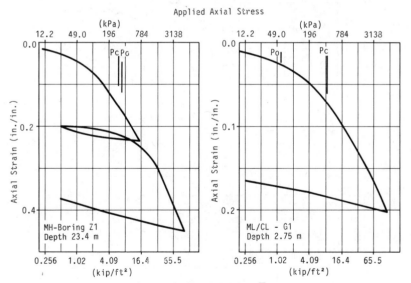

FIGURE 3, SELECTED LABORATORY CONSOLIDATION TEST RESULTS.

The consolidation test results obtained from the samples from borings G3 and D8, which are shown in Fig. 3, are representative of the compressibility of the silty sand stratum. The laboratory recompression ratios, C_{cr}, measured in these tests in a rebound near the maximum expected stresses, are 0.010 and 0.0062. The ratios backfigured from the settlements measured in field test no. 2 are 0.0044 for the center plate and 0.0069 for the corner plate. In this case, the lab value somewhat overestimates the field value, but both indicate a very small compressibility of this stratum under the expected static loads.

The settlement record labeled "outside plate" in Fig. 4 was obtained on a settlement plate located 6.6 ft (2 m) outside the loaded area along its center line. The small movements recorded on this

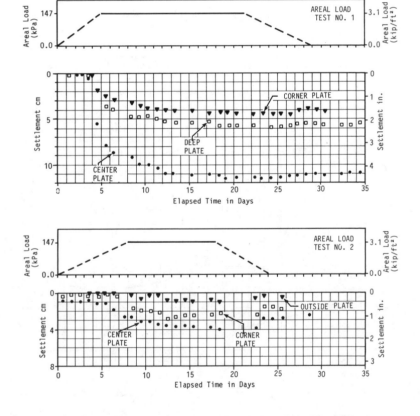

FIGURE 4, RECORDS OF SETTLEMENTS FROM FIELD LOAD TESTS.

plate revealed that the effect of the stockpiling areas on the equipment foundations was going to be negligible. In both field load tests, the coils were piled against two building footings. These footings did not experience any noticeable movements. The piers adjacent to test no. 1 are 79 ft and 121 ft (24 m - 37 m) deep and those near test no. 2 are 95 ft and 112 ft (29 m - 34 m) deep.

CONDITIONS OF THE PIER FOUNDATION

This part of the investigation consisted of drilling 19 borings through the piers to reach the foundation material at their tips and four more adjacent to the piers to explore the soils around the shaft. They were complemented with two pier load tests, the results of which are summarized in Fig. 5. Load test no. 1 was performed on the closest single pier available to the affected footing and test no. 2 was implemented on a short pier embedded mostly in the silty sand stratum.

In 13 of the piers investigated, the borings confirmed that the piers were resting on limestone boulders up to 13 ft (4 m) thick. In the other 6 piers, the borings were stopped after having drilled from 17 ft to 82 ft (5 m - 25 m) in sound limestone.

The core recovery in the concrete was usually larger than 95%. Twenty unconfined compression tests were implemented on specimens cut from the concrete core. Their average strength was 410 Kip/ft^2 (19.6 MPa). The concrete of the piers in the affected footing did not exhibit any defects that could have contributed to the observed settlements.

Pier load test no. 1 was performed on an 87 ft (26.6 m) long pier of 2.8 ft (0.85 m) in diameter. In this test, the settlements shown in Fig. 5 are questionable beyond 760 Kip (3.38 MN), because the concrete on the pier cap was fissured beyond this load. Pier test no. 2 was completed on a 34 ft (10.5 m) long pier with a diameter of 3.5 ft (1.08 m). The characterization of the soil from the results of these two dissimilar tests follows the procedure proposed by Poulos (5). This is based on an elastic analysis of the pile and the embedding material. An overall soil modulus is backfigured from the load test, which can be used then to extrapolate the test results to different pile lengths or diameters, provided that they are founded in similar conditions.

The field load tests indicate that the behavior of the gravel and silty sand layers is fairly similar, and it appears that there is no point in differentiating between the two layers. The soil deposit has been assumed to be homogeneous and about four pile lengths thick in all cases. The elastic modulus of the pier has been estimated (1) to be 131,000 Kip/ft^2 (6.3 MPa) based on the strength measurements on the core specimens.

With these assumptions, the soil modulus is backfigured from the pier settlements corresponding to cap loads of 761 Kip and 684 Kip (3.4 MN - 3.0 MN) for test no. 1 and no. 2, respectively. These loads were chosen considering that the associated settlements were small and, therefore, the possibility of soil slippage along the shaft was small. The overall soil modulus backfigured from test no. 1 was 1090 Kip/ft^2 (52.2 MPa) and from test no. 2 was 890 Kip/ft^2 (42.6 MPa).

The maximum allowable differential settlement between adjacent footings is 0.5 in. (1.2 cm). Considering this as a total allowable settlement and the 66 ft (20 m) depth of the piers in the affected footing, the elastic method indicates that it would take a load of 1047 Kip and 772 Kip (4.65 MN – 3.43 MN) on each pier to cause that

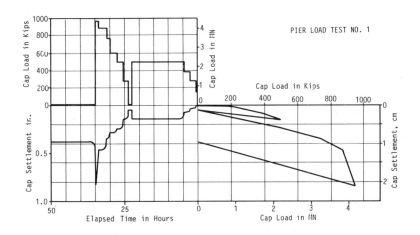

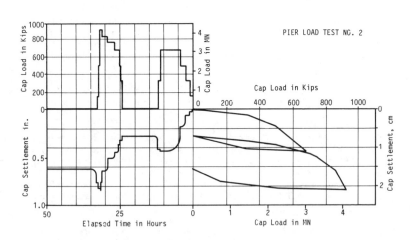

FIGURE 5, SUMMARY OF PIER LOAD TEST RESULTS.

settlement for the soil modulus of test no. 1 and no. 2, respectively. Therefore, the two piers could withstand the static design loads of 473 Kip (2.11 MN) with total settlements smaller than 0.35 in. (0.9 cm).

When the H-pile induced settlements occurred, the applied static loads on the piers were only about one fifth of their design loads. For this load level, the elastic method predicts a total cap settlement of 0.05 in. (0.13 cm) for the modulus obtained from pier load test no. 1 and of 0.07 in. (0.18 cm) for that from pier load test no. 2. These settlements are a negligible fraction of the 10 in. (25 cm) settlement that was observed upon driving the H-piles.

The results of the pier load tests appear to be in general agreement with the results of the field load tests. For comparison purposes, the field load is approximated with a uniformly loaded rectangular area on an elastic medium. Under these conditions, the settlements at the center of the loaded area are calculated (4) for the values of the overall elastic modulus backfigured from the pier load tests. The predicted settlements are 1.75 in. and 2.20 in. (4.5 cm - 5.6 cm) for the modulus from test no. 1 and no. 2, respectively. These settlements are of similar magnitude as those recorded in the field load tests.

POSSIBLE CAUSES OF THE INDUCED SETTLEMENTS

A cross section through the affected footing is shown in Fig. 6. The boring Z4 and Z23 were drilled through the piers of the affected footing and the boring Z1 was drilled adjacent to them. In all borings, the piers appear to be resting on a boulder or ledge of limestone about 3 ft (0.9 m) thick. This boulder is underlain by soils from the silty sand stratum, that in boring Z1 become highly compressible clayey silt, MH. A consolidation test implemented on a sample of this soil from boring Z1 is shown in Fig. 3. The results suggest that the MH is normally consolidated; therefore, it is reasonable to expect that the overlying limestone is a boulder. Nevertheless, the field load tests illustrated how fast the settlements can take place at this site. Since the borings were drilled sometime after the pile driving, it is not clear whether the limestone at the tip of the affected piers was originally a boulder or a rock ledge that had been broken with the H-pile driving.

A possible explanation of the settlements is that they were the result of the interaction of the driven piles and the limestone boulder on which the piers were resting. If the soils underlying this boulder were very loose, the driven piles could have broken or displaced the boulder causing a cave-in like effect. However, this explanation does not seem very likely, since there was no indication on the ground surface that this was happening.

Then the only explanation left is to attribute the settlements to the dynamic compaction induced by the pile driving on the loose to med. dense silty sand layer. Similar cases (2, 3) have been reported. Clough et al. (2) monitored maximum displacements of 6 in. (15 cm) upon driving 50 ft (15 m) long sheet piles. In the present case, the H-piles were also small displacement piles and they were, at least, five times longer. From this comparison, the 10 in. (25 cm) pier settlement observed does not seem excessively large.

In this case, the frequent limestone boulders have probably provided an efficient mechanism to transform the driving energy into the elastic waves responsible for the soil compaction. Clough et al. (2) distinguished one case of "hard driving" when the sheetpiles were obstructed by rubble. Under these conditions, they measured accelerations about twice as large as in normal driving.

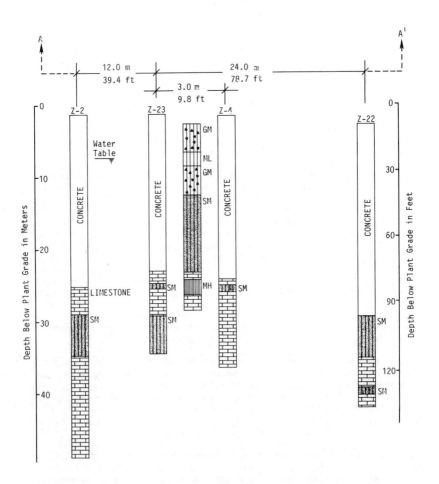

FIGURE 6, CROSS SECTION A-A' THROUGH THE AFFECTED FOOTING

The ground movements induced by pile driving at this site drop from a maximum near the affected footing to zero in the closest unaffected footing located 39 ft (12 m) apart. This drop does not seem excessively large based on the observations of Dalmatov et al. (3). They monitored accelerations at several distances from a driven pile. The subsoil was a silty sand stratum of similar composition to that of this site. There, the peak accelerations dropped below the critical acceleration and therefore, the ground movements were zero at distances beyond 26 ft (8 m) from the pile center.

In summary, although there is no conclusive evidence it appears that the only plausible explanation is to attribute the settlements to the dynamic compaction of the silty sand induced by the H-pile driving. This is based on the fact that the settlements observed are similar to those reported for analogous cases of dynamic soil compaction. Furthermore, the attenuation rate indicated by the fact that the adjacent footings remained unaffected appears to be in agreement with the observed attenuation by Dalmatov et al. (3) for silty sands with similar amounts of fines.

SUMMARY AND CONCLUSIONS

The expansion is located on a predominately granular soil deposit. This consists of a surface layer of med. dense gravel and an underlying stratum of silty sand, from loose to med. dense. The bedrock is limestone and occurs at tremendously variable depths. Very frequently, boulders or ledges of the same type of limestone were found in the silty sand stratum.

Despite their density, both strata exhibited a very low compressibility in two field load tests. The results of two pier load tests indicate that the piers in the affected footing could withstand the static design loads with total settlements smaller than 0.35 in. (0.9 cm).

However, upon driving the steel H-piles near this footing, these two piers settled about 10 in. (25 cm); at that time, the total static load applied on each pier was only about one fifth of the design load. Although there is no conclusive evidence of the cause of the settlements, all available indications point to the dynamic compaction induced by the H-pile driving.

APPENDIX - REFERENCES

1. "Building Code Requirements for Reinforced Concrete," ACI 318-83, American COncrete Institute, Detroit, Michigan, 1983, p. 29.

2. Clough, G.W., and J. L. Chameau, "Measured Effects of Vibratory Sheetpile Driving," Journal of the Geotechnical Engineering Division, ASCE, Vol. 106, No. GT10, Oct., 1980, pp. 1081-1099.

3. Dalmatov, B. I., V. A. Ershov, and E. D. Kovalevsky. "Some Cases of Foundation Settlement in Driving Sheeting and Piles." Proceedings of the International Symposium on Wave Propogation and Dynamic Properties of Earth Materials, University of New Mexico, Albuquerque, New Mexico, pp. 607-613, 1968.

4. Lambe, T. W., and R. V. Whitman, Soil Mechanics, 1st ed., John Wiley and Sons, Inc., New York, 1969, p. 215.

5. Poulos, H. G., "Some Recent Developments in the Theoretical Analysis of Pile Behavior," in Soil Mechanics - New Horizons, Edt. by I.K. Lee, Newnes-Butterworths, London, 1974, pp. 237-279.

VEHICLE INDUCED GROUND MOTION

John A. Barneich*

Introduction

The purpose of this paper is to present typical vibration data
characteristic of vehicular induced ground motions and to review the
conditions that apparently affect the amplitude and attenuation of
ground motion. The sources of data presented represent projects
where vehicular induced vibrations had led to complaints and
litigation by homeowners residing near roadways or where vehicular
induced vibrations were important to the isolation of vibration
sensitive equipment. To meet the needs of this paper and at the same
time keeping the details of a specific project confidential, only
vibration data and general site conditions are discussed. Based on
the available data, emphasis is placed on vibrations due to bus and
truck traffic with some reference to auto and train traffic induced
vibration.

The sections that follow describe: (1) vibration monitoring --
equipment used to make the measurements; (2) background information
on the site conditions, measures of roadway roughness and general
vibration perception limits; (3) ranges of vibration levels
associated with vehicular induced vibrations; (4) the effects of site
conditions and vehicular characteristics on vibration amplitude; and
(5) attenuation of vehicular induced vibrations.

Vibration Measurements

Vibration data described herein were derived from five sites labeled
for discussion as sites A, B, C, D, and E. Vibration data measured
at site A was measured with Electro-Tech Model EV-17 horizontal and
vertical velocity sensitive transducers which were matched for proper
impedance and electrical damping to a CEC Model 5-124 galvo-recorder
with high-sensitivity galvanometers. At sites B, C, and D,
measurements were made using Sprengnether Model S-6000 velocity
sensitive transducers, a Honeywell 117 Accudata multi-channel DC
amplifier and a Honeywell Model 906C twelve channel visicorder. At
site E, measurements were made using Sprengnether Model S-6000
velocity sensitive transducers, a Honeywell 117 Accudata multi-
channel DC amplifier, and a Honeywell 1858 Oscillograph (strip chart
recorder). Secondary vibration measurements were also made at sites

*Vice President and Senior Associate, Woodward-Clyde Consultants, 203
North Golden Circle Drive, Santa Ana, California 92705.

B, C, D, and E using a VS-1600 Sprengnether seismograph. In all cases the transducers consisted of a moving magnetic-coil system which transforms mechanical vibration into electrical signals. The primary measurement systems were generally able to sense peak-to-peak particle velocities in the range of 0.0001 inches per second (ips) to 3 ips at frequencies from 2 to 200 Hertz (Hz). All measurements were made using two sets of transducers to obtain simultaneous measurements at two locations. In general, the location of one set of transducers was fixed at a reference location while the second set was moved to various locations of interest.

Background Information

Because of project constraints at the five sites, only general subsurface soil conditions were known. These conditions are summarized for each site in Table 1. The roadway conditions at sites A through D consisted of an asphalt road surface with concrete curbs and gutters and adjacent concrete flat work. The concrete flat work adjacent to streets at sites A and B was well developed with connecting driveways, sidewalks as well as underground utilities. Site C was well developed without sidewalks and more exposed soil. Site D was undeveloped with no sidewalks or driveways. The conditions at site E were undeveloped with the land adjacent to the railroad and roadway (freeway) consisting of dirt roads and undeveloped land.

TABLE 1

Summary of Site Subsurface Conditions

Site	General Subsurface Soil Conditions
A	Soft to medium stiff clay in top 10 feet underlain by medium dense to dense clayey sand and silty sands. Water not encountered in top 20 feet.
B	Loose to medium dense sands and silt. Water table within top 10 feet.
C	Medium stiff to stiff clayey soils. Water table greater than 10 feet.
D	Medium stiff to stiff sandy clays and medium dense clayey sands. Water table greater than 40 feet.
E	Medium stiff to stiff sandy clays and medium dense clayey sands. Water table greater than 40 feet.

The natural roadway surfaces at sites A, B, D, and E were relatively smooth, whereas at site C the roadway was moderately rough. The roadway at sites A, B, and D was made artificially rough for selected measurements by constructing low asphalt berms perpendicular to traffic or the placement of 2" x 6" and 2" x 12" planks of wood across the roadway perpendicular to traffic. A measure of roadway roughness equal to bump height divided by the square root of the wave length of the bump was used in analyzing data. A summary of these conditions for each site is presented in Table 2.

For reference purposes, the studies at site A were completed in response to pending litigation where it has been alleged that vibration caused by a regularly scheduled bus running over a freshly constructed asphalt berm (overbuilt covering a utility pipe trench crossing the street) had caused distress to several adjacent residences. Studies were completed at sites B, C, and D to respond to complaints of buses causing disturbing vibration levels at residences adjacent to heavily travelled bus routes. Studies at site E were conducted to develop base level vibration data from train traffic to a proposed technology center.

For reference purposes, various levels of "steady state vibration" perception were chosen from the literature (1) (2) and are plotted in Figure 1 (i.e., not noticeable to persons, barely noticeable to persons, etc.). It is the writers experience that the vibration levels identified in Figure 1 are also generally reasonable for "transient vibrations" of the type induced by vehicular traffic. It is noted, however, that the effect of vibration levels on persons varies from person to person, and depends on the environment and the activity that the observer is engaged in when vibrations occur. For example, in a quiet environment with the observer sitting or laying down, he or she will be least sensitive to vibration. Also, some persons are more sensitive to vibration than others. Similar limitations apply to the vibration tolerance levels of machines and structures shown in Figure 1.

Characteristic Vehicle Induced Vibration Levels

A summary of vibration levels monitored at sites A, B, C, D, and E due to auto, bus, and truck traffic is presented in Table 3. The vibration levels in Table 3 generally represent peak particle velocity and dominant frequency monitored for the identified traffic condition. Also, these are generally representative of vertical vibration amplitude with horizontal vibration amplitudes being generally one-half to two-thirds of vertical amplitude in the same frequency range. The weight, wheel base, and speed of the vehicle responsible for the vibrations together with the roadway conditions (type of surface and roughness) and site identification are also summarized in Table 3. Also, the vibration data shown in Table 3 were monitored at the side of the roadway 10 to 20 feet from the vehicle travel path. Other data monitored at greater distances are not shown in Table 3 but will be discussed in later sections.

TABLE 2

Summary of Roadway Roughness

Site	General Roadway Condition	Approximate Roadway Roughness $R(in^{1/2})$*
A	Smooth asphalt surface	< 0.03
A	Low artificial asphalt berm characterizing average asphalt overfill over utility trench	0.2 - 0.35
A	High artificial asphalt berm characterizing maximum asphalt overfill over utility trench	0.3 - 0.5
B	Smooth asphalt surface	< 0.03
B	Artificial bump constructed with 2 - 2 x 12 wood planks** side by side and a 2 x 6 wood plank on top across the roadway	0.5 - 0.6
C	Moderately rough asphalt surface	0.06 to 0.15
D	Smooth asphalt surface	< 0.03
D	Artificial bump constructed with a 2 x 6 wood plank across the roadway	0.4 - 0.5
E	Smooth asphalt surface	< 0.03

----

* Roughness: $R = \dfrac{\Delta}{\sqrt{L}}$

** All wood planks were smooth cut bumper and attached to the roadway with Ramset nails.

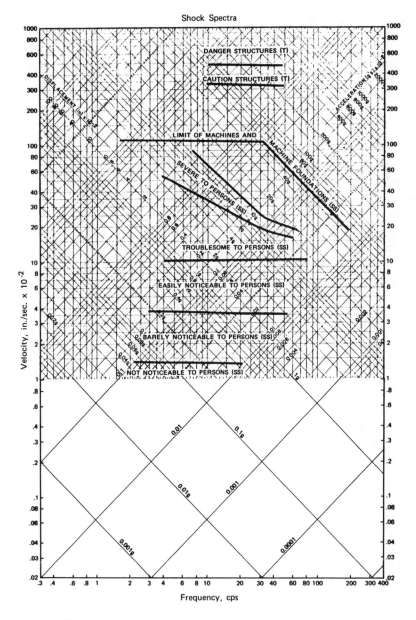

Figure 1 – Reference Levels for Vibration Perception

TABLE 3

Summary of Auto, Truck and Bus
Indirect Vibration Levels at Road Site
Location 10 to 20 feet from the Vehicle

Vibration Level & Frequency		Vehicle			Speed	Road Conditions		
Peak Particle Velocity (ips)	Frequency (Hz)	Weight* (kips)	Type	Wheel Base (ft)	(mph)	Road Surface	Roughness	Site
.20 - .33	9 - 17	19	Bus	24	45 - 49	Asphalt	.2 - .35	A
.15 - .22	12 - 15	29	Truck	14	41 - 45	Asphalt	.2 - .35	A
.32 - .36	11 - 15	29	Truck	14	39 - 44	Asphalt	.3 - .5	A
.001 - .009	3 - 14	?	Autos	?	20 - 35	Asphalt	< .03	B
.01 - .015	10 - 13	?	Truck	?	25 - 35	Asphalt	< .03	B
.001 - .017	13 - 20	14	Bus	11.4	10 - 45	Asphalt	< .03	B
.001 - .028	11 - 14	22	Bus	25	10 - 35	Asphalt	< .03	B
.003 - .02	10 - 12	27	Bus	25	10 - 35	Asphalt	< .03	B
.09 - .20	7 - 10	22 - 27	Bus	25	20	Asphalt w/addt'l bump		B
.0015 - .15	10 - 30	?	Autos	?	25 - 35	Asphalt	.06 - .15	C
.01 - .02	15 - 35	?	Truck	?	25 - 35	Asphalt	.06 - .15	C
.002 - .05	10 - 21	14	Bus	11.4	25 - 50	Asphalt	.06 - .15	C
.005 - .06	10 - 16	22	Bus	25	25 - 50	Asphalt	.06 - .15	C
.003 - .05	12 - 15	27	Bus	25	25 - 50	Asphalt	.06 - .15	C
.001 - .003	20	?	Auto	?	15 - 40	Asphalt	< .03	D
.009 - .015	12 - 17	?	Auto	?	20 - 30	Asphalt w/artificial wood bump	.4 - .5	D
.001 - .013	11 - 27	22 - 27	Bus	25	20 - 60	Asphalt	< .03	D
.04 - .09	10 - 20	22 - 27	Bus	25	20 - 40	Asphalt w/artificial wood bump	0.4 - 0.5	D
.001 - .012	20 - 35	?	Truck	?	30 - 50	Asphalt	< .03	D
.001 - .004	30+	?	Auto	?	50 - 60	Asphalt	< .03	E
.005 - .009	13 - 30	?	Truck	?	50 - 60	Asphalt	< .03	E

----

* total vehicle weight

A summary of vibration levels monitored at site E due to train traffic is presented in Table 4. In contrast to the data in Table 3, vibration amplitude in Table 4 represent measurements made in all of the three orthogonal directions (range of peak amplitude in each direction reported) and are presented for various distant ranges from the railroad track. Further, the ratio of horizontal longitudinal and transverse vibration amplitudes to vertical vibration amplitudes was generally in the range of 0.7 to 1.7 with a mean value of about 1.2. The distances of the measurements from the railroad track and the speed of the train are also presented in Table 4. In all cases the train was a high speed commuter train. Limited measurements of a slow moving freight train (20 to 30 mph) at site E indicated vibration amplitudes in the range of 20% to 50% of those shown for the high speed commuter trains.

TABLE 4

Summary of Train Traffic
Vibration Levels Site E

Vibration Level and Frequency		Train Speed	Distance From Truck	
V (ips)	F (Hz)	(mph)	(ft)	Site*
.05 - .12	10 - 23	40 - 60	50	E-1
.02 - .058	9 - 24	40 - 60	100 - 150	E-1
.007 - .035	9 - 20	40 - 60	200 - 300	E-1
.04 - .10	8 - 26	40 - 60	50	E-2
.01- .035	6 - 25	40 - 60	100 - 150	E-2
.004 - .016	6 - 25	40 - 60	200 - 300	E-2
.014 - .10	10 - 40	40 - 60	30	E-3
.007 - .055	9 - 28	40 - 60	80 - 90	E-3
.005 - .017	7 - 25	40 - 60	130 - 230	E-3

* E-1, E-2, and E-3 represent different locations at Site E with apparently similar site surface and subsurface conditions.

The ranges of vibration readings summarized in Tables 3 and 4 are plotted with respect to amplitude and frequency in Figures 2 and 3, respectively. These data are discussed further in the paragraphs that follow.

Affects of Site Condition and Vehicle Characteristics on Vibration Amplitudes

As can be seen in Figure 2, auto, bus, and truck traffic induced vibration levels at the curb side range from below the perception range of persons up to troublesome and severe to persons covering vibration amplitudes from 0.0001 to 0.35 inches per second (ips) peak particle velocity. Frequencies are generally in the 3 to 30 Hertz (Hz) range with most data in the range of 7 to 20 Hz. As can be noticed in Figure 2, truck and buses travelling over rough roadways cause maximum vibrations with the vibration levels reducing with reducing roadway roughness. Further, the vibration levels of trucks and buses over smooth roadways is approximately equivalent to vibrations caused by auto traffic travelling over a rough roadway surface. This effect is more specifically shown for bus traffic at site B in Figure 4. Basically, Figure 4 shows the variation of vibration amplitude (peak particle velocity) versus roadway roughness for three types of buses. Though there is considerable data scatter, Figure 4 indicates an approximately linear increase in vibration amplitude with roadway roughness with the roadway surface having almost twice the effect on the heavier, larger buses (B and C) than for the small bus (A). It is further noted that for bus B the increases in vibration level begins at a higher rate than bus C but at high levels of roughness the vibration levels from bus C are increasing at a higher rate than bus B. These trends are likely related to the tuning of the suspension of the buses, to the speed of the bus, and roughness of the roadway. Insufficient data were obtained to further explore this phenomenon.

The effect of the speed of the bus on increased vibration levels was also considered based on the data from sites B and C as shown in Figures 5 and 6. As can be seen from Figures 5 and 6, the vibration amplitudes increase almost linearly with vehicle speed for each of the bus types. The data in Figure 5 are representative of smooth roadway conditions (R < 0.03) at site B while in Figure 6 the data are representative of moderately rough roadway conditions (R = 0.06 to 0.15 in$^{1/2}$).

Vibration Attenuation with Distance

The attenuation of vibration amplitude with distance has been represented by (3) the following semi-empirical relationship:

Shock Spectra

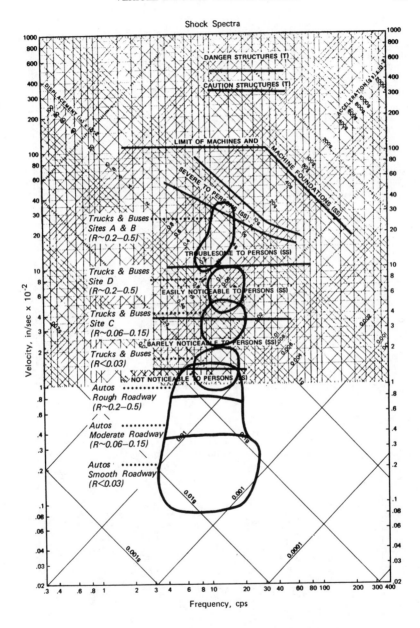

Figure 2 - Summary of Curb Side Vibration Levels
Autos, Trucks, and Buses

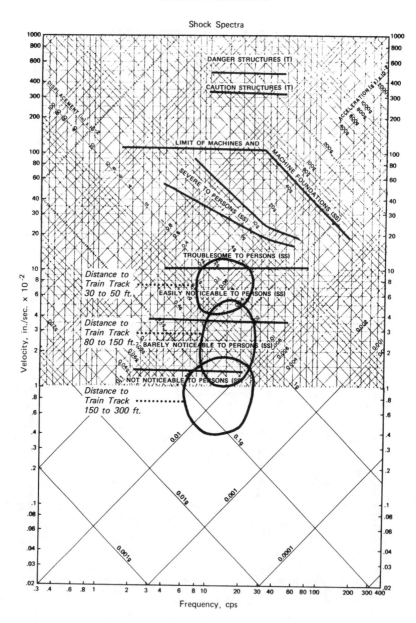

Figure 3 - Summary of Train Induced Vibration Levels

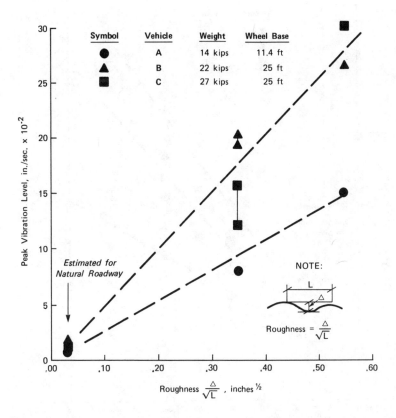

Figure 4 - Effect of Roadway Roughness on Vibration Amplitude (Site B)

VIBRATION PROBLEMS

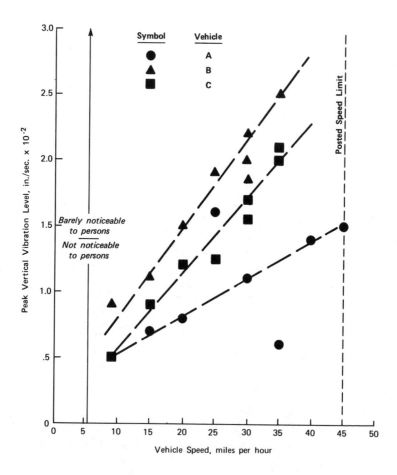

Figure 5 - Effect of Vehicle Speed on Bus Induced Vibrations (Site B)

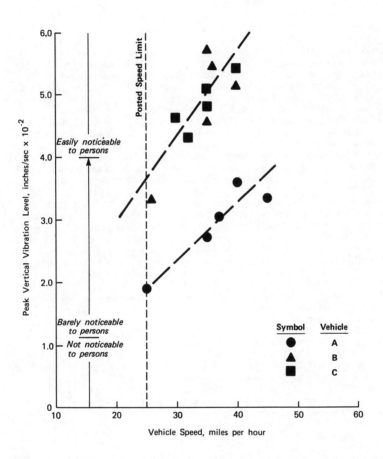

Figure 6 – Effect of Vehicle Speed on Bus Induced Vibrations (Site C)

$$\frac{A_2}{A_1} = \sqrt{\frac{r_1}{r_2}} \ e^{\alpha \ (r_1 - r_2)}$$

where: A_1 = amplitude of vibration at level 1
        A_2 = amplitude of vibration at level 2
        r_1 = distance of location 1 from the source in feet
        r_2 = distance of location 2 from the source in feet
        α = attenuation coefficient in feet^{-1}

It noted that the square root term is to account for attenuation due to geometric spreading and the exponential term is to account for the hysteretic damping of the soil. Further, the attenuation coefficient " α " is frequency dependent. Also, from the form of the equation, the attenuation increases with increasing " α ". The vibration data from sites B, C, D, and E were analyzed with respect to the above attenuation equation with the resulting attenuation curves fitting the mean values of " α " for each site presented in Figure 7. Also shown in Figure 7 is the expected attenuation due to geometric spreading and an indication of data scatter for each curve represented by the vertical range arrows shown on each curve. The scatter of data is considerable for curves for sites B and C, slightly less for site D, and considerably less for site E. In fact, as seen in Figure 7, some of the data from sites B and C were very close to or above the expected attenuation curve with respect to geometric spreading. This and the large scatter was probably due in part to the highly developed concrete flat work and utilities that existed at sites B and C. The lower data scatter for sites D and E is probably due in part to the progressively lower development of concrete flat work and utilities. Another contribution to the data scatter is likely the range of frequencies represented by the data varying by about a factor of $\pm$ 50 to 75% about a mean value. Though the attenuation coefficient has been related to soil type (1) (2) (3), with the more soft or loose soils exhibiting higher attenuation coefficients (causing more attenuation) than stiff or hard soils, it appears from the examples shown in Figure 7 man-made improvements may dominate the attenuation in some instances. For example, site B exhibited the soft test soil conditions of the three sites, yet its mean attenuation coefficient was lower than either site C or D. Further, it is likely that data from site E (with no significant man-made improvements) reflect most nearly the attenuation characteristics of the subsurface soils.

Acknowledgements

The writer wishes to acknowledge the involvement of Dr. R. L. McNeill in the work completed on site A, Mr. B. Bevier for his assistance in vibration measurements on sites B, C, D, and E, and Dr. P. Ramsamooj for his assistance in vibration measurements made at sites B, C, and D.

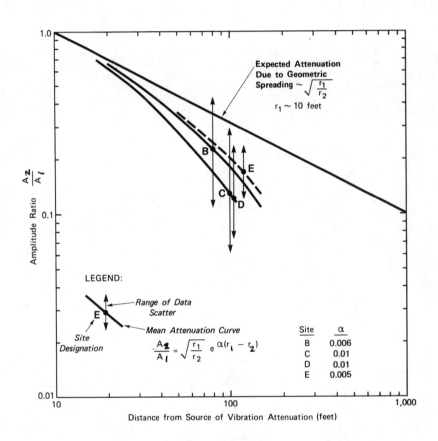

Figure 7 - Summary of Vibration Attenuation

VIBRATION PROBLEMS

References

(1) McNeill, R. L., "Machine Foundations--The State of the Art," Proc. 7th International Conference on Soil Mechanics and Foundation Engineering, Mexico City, Mexico, August, 1969.

(2) Richart, F. E., Hall, J. R., and Woods, R. D., "Vibrations of Soils and Foundations" Prentice Hall, Inc., Englewood Cliffs, New Jersey, 1969.

(3) Barkan, D. D., "Dynamics of Bases and Foundations", (translated from the Russian by L. Drashevska, and translation edited by G. P. Tscheborarioff), McGraw-Hill Book Co. (New York), 434 pp., 1962.

OBSERVED HIGH-RISE BUILDING RESPONSE TO
CONSTRUCTION BLAST VIBRATIONS

Lewis L. Oriard, M. ASCE[1]
Thomas L. Richardson, A. M. ASCE[2]
Kenneth P. Akins, Jr., M. ASCE[3]

ABSTRACT: Five case studies of vibration monitoring of multiple-story buildings during construction of the Metropolitan Atlanta Rapid Transit Authority's North Line are presented. Conventional drill and blast techniques were used adejacent to, and in some cases underneath, low-rise and high rise structures. Vibration monitoring instruments were placed in the upper levels of several structures adjacent to the construction. Monitoring instruments were also placed in the basement of the structure to measure the blast induced vibrations input to the structure. The data gained yielded valuable case study information on the response of structures to blast induced vibrations. The intensity and frequency of the vibrations at the selected locations of the structures are compared to the input vibrations. Typical acceleration, displacement, velocity, and frequency information obtained are presented. The data indicated that the response of the structures varied. In most cases a decrease in the intensity of the vibrations was observed, with an increase in intensity occurring in upper levels when compared to intermediate levels for some cases.

INTRODUCTION

The construction of the Peachtree Center Station and adjoining tunnels of the Metropolitan Atlanta Rapid Transit Authority's (MARTA) "North Line" required considerable underground excavation in the heart of Atlanta's central business district. The excavation was accomplished by conventional drilling and blasting techniques adjacent to, and

---

[1]President, Lewis L. Oriard, Inc. Huntington Beach, CA
[2]Project Engineer, Law/Geoconsult International, 1140 Hammond Drive, Atlanta, Georgia, 30328
[3]Senior Geotechnical Engineer, Law Engineering Testing Co., P. O. Box 21879, Columbia, South Carolina, 29221

in some cases underneath, low-rise and high-rise structures.
As part of the quality assurance procedures for the project,
MARTA commissioned an extensive program of ground vibration
and air blast monitoring. The response of the multi-story
structures to the ground vibrations was of particular
interest. Therefore, during some of the blasting operations, monitoring instruments were placed in the upper
levels of several high-rise structures adjacent to the
construction.

The primary purpose of this paper is to examine the
results of field monitoring of the responses of high-rise
buildings to small-scale blasting operations nearby, as
illustated in five case studies. Important concepts not
always recognized about the response to this vibration
source are emphasized. Information is provided on the site
characteristics, types of blasting and structures where
vibrations were monitored. The trend observed in the data
and supplemental findings of the study are applicable to
future projects.

SITE DESCRIPTION

     , The site of the construction activity was in the heart
of downtown Atlanta, Georgia. The dominant topographic
feature is a north-south ridge. Peachtree Street follows
the ridge line, and numerous low-rise and high-rise structures are located on either side of the street.

The rock in the ridge is typically gneiss of good to
excellent quality. Rock cores obtained during the subsurface investigation for the work typically had core recover-
ies in excess of 90 percent and rock quality designations
(RQD's) typically over 80 percent. The Peachtree Center
Station is approximately 100 feet below the ground surface
in a rock chamber excavated in the ridge.

Line tunnels extending to the north of the station are
in rock of similar quality, although poorer conditions are
present at the northern limit. Line tunnels to the south of
the station are in rock for about 900 feet (275m). Beyond
that point, there is a gradual transition from a rock
section, to mixed faced conditions, and then to soft ground
conditions. The softer materials are loose to firm
micaceous silty sands derived from in-place weathering of
the parent gneisses. Multi-story structures near this
portion of the tunnel construction typically have
foundations bearing on these residual soils. High-rise
structures are supported by deep foundations extending near
or into the underlying rock.

DESCRIPTION OF BLASTING

The Contractor used conventional drilling and blasting
techniques to fragment the rock to facilitate its removal.
Water gel blasting agents were used throughout the work.

Both millisecond and long period delays were used, the choice depending upon the type of blasting. Blasting patterns included those for full-face rounds, slashing rounds, bench excavations, vertical and inclined shaft sinking and raising, trench excavation and trim shooting. As much as 800 pounds (3.56 kN) of the explosives were detonated in a single blast. However, the maximum weight of explosives detonated in a single delay was reported as not exceeding 45 pounds (0.2 kN).

MONITORING EQUIPMENT

Before construction work began, the senior author was engaged in various discussions with the Project Owner, Engineer, and Consultants regarding planning, developing criteria, writing specifications and making the related decisions with respect to the blasting. One of the decisions made was to use particle velocity criteria to limit blasting vibrations, and to use monitoring instruments which would respond directly to particle velocity. This decision was based on the widely accepted concept that particle velocity is more directly related to a structure's response to blasting vibrations than either acceleration or displacement (the other parameters of particle motion readily measured, also).

For an in-depth study of a specific structure, it would be desirable to measure strains induced by the ground motion. However, this is generally not necessary, nor is it generally an acceptable approach to the routine monitoring of an urban blasting project. When a structural response is monitored, there is normally a demand that the work be done unobtrusively, with no defacing of the structure and no interference with the activity of its occupants. These demands can be met with portable transducers placed on suitable floor positions, but not with strain gauges bonded to walls or structural members.

The monitoring equipment consisted of two types of particle velocity recorders. One type of recorder was self-triggering, which recorded vibrations above a pre-selected threshold. The record of the event provided a trace of the components of particle velocity versus time, allowing frequency determination and making possible further analysis of the motion. The other type of recorder registered only the maximum component of particle velocity, and did not allow frequency resolution. Both instruments had a linear response for vibration frequencies from 1 to 200 Hz. The instruments could also respond to vibrations with frequencies outside the linear response range, so that structural motions of even lower frequencies would have been detected.

Recorders were placed on the lowest basement floor level to monitor incoming vibrations to gather information on the response of multi-story structures. Additional recording was then done on the upper floors. Transducers were oriented so that one horizontal axis was parallel and one perpendicular to the tunnel alignment at the location. These axes coincided with building axes with the exception of one building with a triangular floor plan.

DATA PRESENTATION AND ANALYSIS

The objective of this study was to compare base and upper level vibrations induced by blasting for a variety of structures. Comparisons were made in three general ways:
1. Idealized response spectra envelopes;
2. Equivalent scaled distance relationship;
3. Event-by-event comparison of amplitudes.

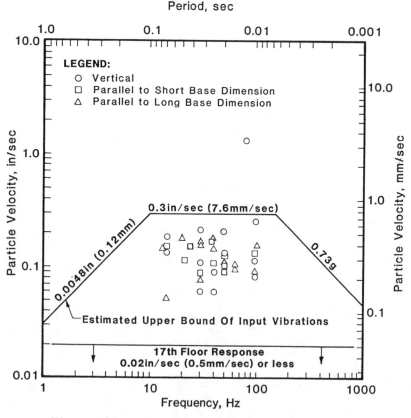

Fig. 1—Observed Input Vibrations
and Response of Building A

Idealized Response Spectra Envelopes

Neither the original data form nor the depth of this study seemed to warrant development of actual velocity spectra. However, in order to present the data in a form which would provide a simple visual comparison of spectral forms, envelopes of idealized spectra (as in Fig. 1) were prepared as follows:

1. The maximum particle velocity for each component of motion was plotted for each blast as a function of frequency;

2. An upper bound line was visually estimated for particle velocity, and extended to a frequency of 2/3 that of the lowest frequency data point and to 1-1/2 times the frequency that of the highest frequency data point;

3. The response was then assumed to drop off at 45 degrees (equivalent to constant displacement and constant acceleration for left and right sides, respectively).

This approach was judged by the authors to be suitable for the desired comparison, on the basis of prior experience. The resulting shape of the response spectra was similar to that resulting from analytical and experimental work by Hendron (1977) who also cited the work of several others; Hudson and Housner (1955); Naik (1979); and Dowding and Corser (1980).

Equivalent Scaled Distance

Researchers and practitioners routinely plot observed peak particle velocities as a function of "scaled distance" (a function of the ratio of distance and weight of explosives). The mathematical form of the Oriard (1972, 1980) upper bound for downhole blasting with normal confinement may be written:

$$v = H \left\{ \frac{d}{w^{1/2}} \right\}^{-1.6} k_1, k_2, k_3 \ldots$$

in which H = a site constant having a typical upper bound value of 242 in average rock; v = peak particle velocity; d = distance; w = charge weight of explosive per delay; and k_1, k_2, k_3, ... are dependent on various blasting design parameters. The combined value of k factors equals 1.0 as a typical upper bound for unconfined blasting. The equation is empirical; therefore, consistent units, English or metric, and the associated constant must be used. The suitability of the Oriard scaling law was confirmed for groundborne vibrations by records for the over 3600 detonations of explosives on the project.

To present a simple demonstration of vibration attenuation within the buildings studied, the concept of "equivalent scaled distance" was developed. The distance from the blast to the nearest point on the base of the structure was added to the distance the vibrations traveled through the building to obtain the total travel distance of the vibrations. This total distance was then divided by the square root of the weight of explosives per delay to obtain the "equivalent scaled distance".

Event by Event Amplitude Comparison

Scaled distance plots are useful in representing overall attenuation patterns. However, the overlap in range of observed amplitudes did not allow an effective comparison of upper level and base level vibration. Therefore, comparisons on an event-by-event basis were made.

CASE STUDIES

The five structures selected for monitoring vibrations effects with recorders at multiple locations in the building were chosen to provide a range of building age and type, as can be seen in Table I. Building A is similar to several other structures adjacent to the tunneling operations. Building B represents a high-rise concrete frame whose height is comparable to its least base dimensions, while Building C is a slender reinforced concrete tower. Buildings D and E represent older steel frame structures. Building E is known to have a jack-arch floor similar to other old structures in the area.

Building A

To consider the nature of vibrations at the base of Building A, recorded ground motions were used to develop the envelopes of idealized spectra of Fig. 1. With few exceptions, the components of particle velocity were in the range of 0.09 to 0.25 in/sec (2.0 to 6.4 mm/sec), with frequencies between 15 and 100 Hz. During the time these vibrations were experienced by the lower portion of the building, a recorder was in place on the seventeenth floor. Vibrations at that position were not sufficient to "trigger" the instrument and, therefore, were less than 0.02 in/sec (0.5 mm/sec). This lack of response indicates, on the average, that vibrations were attenuated through the building frame to about 1/3 to 2/3 of the input velocity or less.

The highest value on the graph, 1.34 in/sec (34.0 mm/sec), shows the vertical component from a blast detonated almost directly under, but 100 feet (30m) below, the recorder location. No response was detected on the seventeenth floor for this blast.

Analysis of the vibration records for ground motions indicated that for half of the blasts recorded, the peak

TABEL I - Description of Buildings Selected for Multi-level Monitoring

BUILDING	BASE DIMENSIONS*		FLOOR LEVELS		CONSTRUCTION
	Parallel to tunnel	Perpendicular to tunnel	Below Ground	Above Ground	Frame; Foundation Type; Year of Construction
A	90 feet (27m)	240 feet (73m)	2	24	Steel; Shallow drilled piers to rock; 1969
B	190 feet (58m)	380 feet (116m)	1	22	Reinforced Concrete; Shallow drilled piers to rock; 1962, 1968
C	140 feet (43m) Tower diameter is 120 feet (36m)	400 feet (122m)	5 4 5	6 8	Reinforced concrete; Shallow foundations on rock; 1975
D	120 feet (36m)	10 feet to 55 feet (3m to 18m)	1	11	Steel; not known; 1897
E	85 feet (26m)	175 feet (53m)	1	10	Steel; Shallow foundations on soil; 1901

*For Buildings A, B, C, and E, the short base dimension is parallel to the tunnel.
For Building D, the long base dimension is parallel to the tunnel.

particle velocity for the vertical component was the largest of three components. The velocity component parallel to the long axis of building (perpendicular to the tunnel) was the largest in one-third of the cases. For the remainder of the cases, the maximum velocity component of incoming vibrations was parallel to the short axis of the building.

Building B

Input vibrations from blasting were recorded, at various times, at three locations of the lowest floor level of Building B, a 22-story reinforced concrete structure. The ground response spectra of the vertical and horizontal components of particle velocity versus frequency was prepared, utilizing data from all three locations (see Fig. 2).

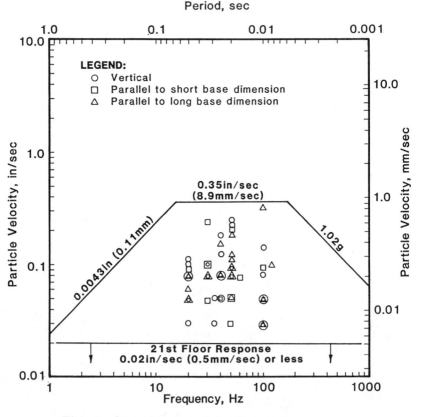

Fig. 2—Observed Input Vibrations
and Response of Building B

The components of particle velocity were between 0.03 in/sec (0.8 mm/sec) and 0.35 in/sec (8.9 mm/sec) with frequencies between 20 and 120 Hz. A recorder in place on the twenty-first floor during the same time recorded no peak particle velocities greater than 0.02 in/sec (0.51 mm/sec), thereby indicating a reduction of at least one-third below the input value. For many of the blasts for which effects were recorded at the lower building level, the device at the upper building level was not triggered, thus indicating even lower peak particle velocities and greater attenuation.

An associated observation was that the upper building level location was subject to considerable background vibrations of 0.02 in/sec (0.51 mm/sec) or slightly greater, but which were not coincident with the time of blast events.

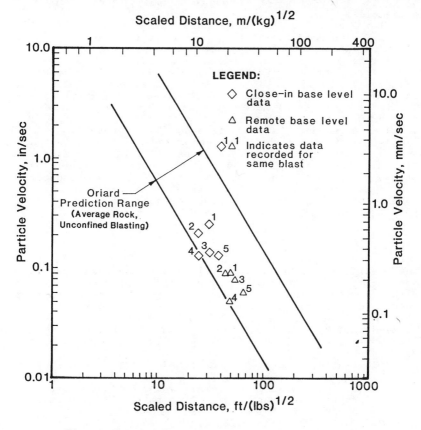

Fig. 3-Observed Attenuation
Across Base of Building B

Conversely, a spurious vibration of 0.02 in/sec (0.51 mm/sec) happening in the same minute may have been attributed inadvertently to a blast.   In any case, peak particle velocites at the upper building level were consistently lower than those at the lower building level (see Fig. 2).   The recordings of vibrations of small amplitude also provide insight in two areas:
 1. Continuous monitoring capability of the equipment;
 2. The amplitude of ambient vibrations relative to incoming vibrations induced by blasting.

   For Building B, the highest value of particle velocity was the vertical component in just over half the records examined.   In the remainder of the records, the highest particle velocities were roughly equally divided into 3 categories:
 1. Parallel to the short base dimension,
 2. Perpendicular to the short base dimension, or
 3. Two or more components essentially equal.

   Recorders were in place at more than one location on the base of Building B for a number of events.   Therefore, an examination of the data for possible attenuation (a reduction in amplitude) across the base was made.   The events for which two recording stations recorded a response are illustrated using a "scaled distance plot" (Fig. 3).
   Only data for which two recording stations detected a peak particle velocity above a threshold value were used for Fig. 3.   Additional data was available for which the closer of two recording points recorded a peak particle velocity and the other did not.   Examination of this data corroborated the general appropriateness of the exponent (-1.6) of the Oriard predition range for average rock.

Building C

   The third building considered involves the most interesting and most complicated case of building response monitoring.   The building has a reinforced concrete frame with a 4-story above ground low-rise portion, with basement levels for parking extending 5 levels (roughly 60 Ft.) below the ground surface along Peachtree Street.   A 120-Ft. (36m)-diameter, 700-Ft. (213m) reinforced concrete tower is located roughly 225 Ft. (69m) away from the street (and MARTA station).
   Initially, recorders were placed at three locations in the structure:
 1. On the lowest floor level in the low-rise portion of the structure adjacent to the MARTA main cavern excavation ("close-in base recorder");
 2. On the fifteenth floor of the tower;
 3. On the seventy-third floor of the tower.

Later during the monitoring, a fourth recorder was posi-
tioned on the lowest floor level of the low-rise portion 400
feet (122m) from the MARTA tunneling.
      For the close-in base recorder, only peak readings of
particle velocity were recorded. Frequency and component
resolutions are not available. However, the type of
blasting and its location relative to the structure were
similar to that for Buildings A and B. Therefore, it is
reasonable to assume that the frequencies and component
make-up of incoming vibrations were similar to those
described for Buildings A and B and illustrated by Figs. 1
and 2.
      The typical frequencies for the fifteenth and
seventy-third floor varied from 10 to 120 Hz (see Fig. 4).

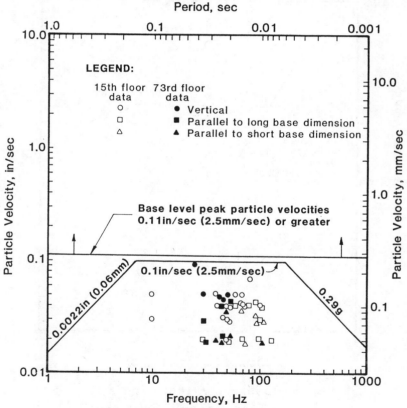

Fig. 4–Observed Input and Response,
          Building C

For the recorded vibrations on the fifteenth floor, the vertical component was the largest component in 12 of the 18 cases. For the seventy-third floor, the vertical component was larger than the other components in four of five cases. In the fifth case, it was equal to the component perpendicular to the tunnel axis.

Additional monitoring at the northwest corner of the building indicated some attenuation of vibration effects across the base just as in the case of Building B (see Fig. 5).

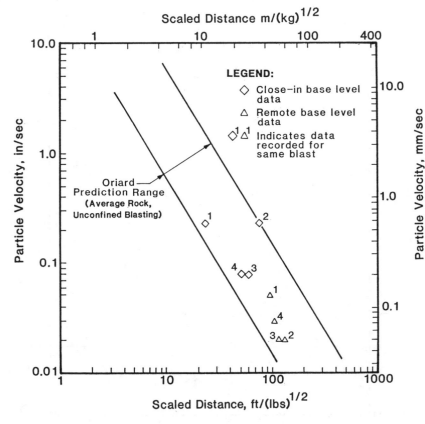

Fig. 5–Observed Attenuation
Across Base of Building C

In view of the attenuation across the base, and the location of the tower farther from the blasting than the close-in base recorder, the peak particle velocity input to the base of the tower was of interest. The data were thus examined as a group on an equivalent scaled distance plot (Fig. 6) with the estimated range of velocities at the base shown.

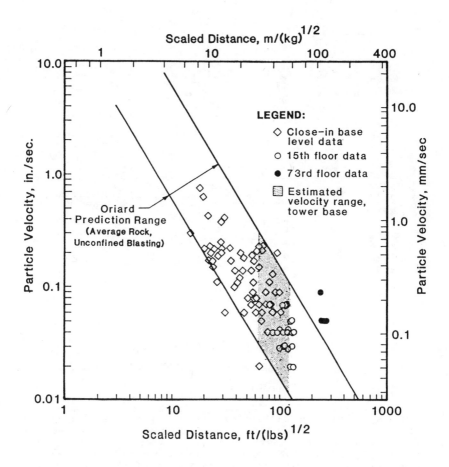

Fig. 6.- Scaled Distance Comparison
of Base and Upper Level Data,
Building C

A comparison on an event-by-event basis was also made (Fig. 7). The vibrations at the tower base were no doubt less than those at the close-in recorder location. Therefore, the estimated range of vibrations at the tower base is also provided on Fig. 7.

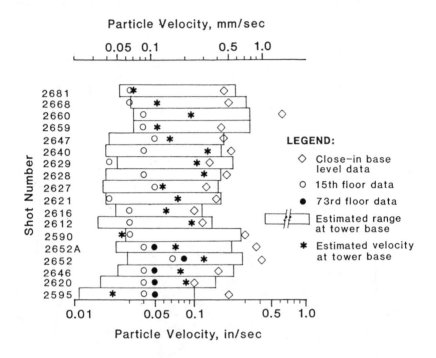

Fig. 7–Shot–by–shot Comparison
of Base and Upper Level Data,
Building C

A specific prediction of the peak particle velocity at the tower base was made. The relative position of the observed particle velocity at the close-in recorder in its predicted range was used to estimate the value at the tower base in its expected range. Much of the range in the scaling is due to the variability from one shot to the next. Experience has shown that multiple observations of a given shot normally occupy the same relative positions in their predicted ranges. Therefore, a better prediction for a single shot is possible if at least one observation at another location has been made.

Building D

Building D is the oldest building selected for monitoring. The 11-story building is quite narrow, and triangular in plan. Monitoring devices were placed in the basement and on the sixth floor. No frequency and component information was obtained at the basement location; however, the information from other ground level recorders in the area indicated no component was consistently largest. The range of vibration frequencies and components measured on the sixth floor is illustrated by Fig. 8.

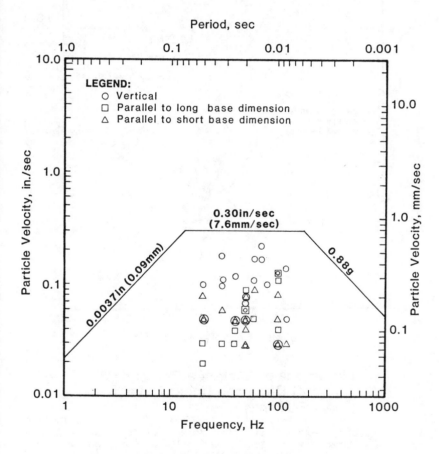

Fig. 8 – Observed Vibration Response,
Sixth Floor of Building D

Peak particle velocities measured in the basement and on the sixth floor are shown in the form of a scaled distance plot in Fig. 9. As in the previous case, the data generally fits within the expected ground motion bounds.

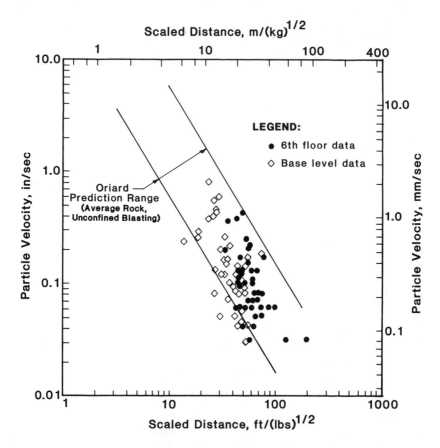

Fig. 9-Scaled Distance Comparison of Base and Upper Level Data, Building D

A peculiarity exists in the data due to the relative recorder locations. The sixth floor location was not directly above the basement location, and, in some cases, the sixth floor location was closer to the blast than the basement location. This positioning contributes to the overlapping of the upper and lower level data points caused by combining the data from all shots on one figure.

Comparison of data on an event by event basis is shown on Fig. 10. The probable range of peak particle velocity at a

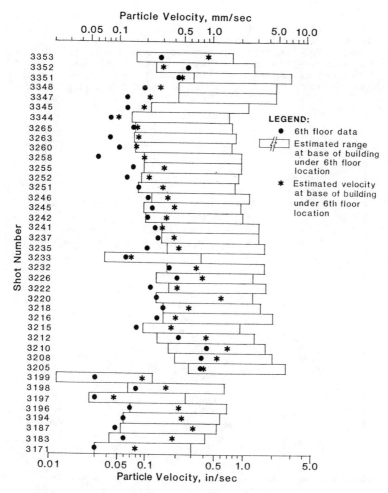

Fig. 10—Shot-by-shot Comparison
of Base and Upper Level Data
Building D

point at the base of the building directly beneath the sixth floor recorder has been estimated. In addition, the most probable peak particle velocity for each shot at that point of building base was estimated by using other base level recorder data and applying the scaling relationship as was done for the Building C data.

Building E

Data was obtained by recorders located on the fifth and tenth floors of Building E, a ten-story steel frame structure. The recorder placed in the basement did not function during this time period. Data from the fifth and tenth floors is shown as an idealized response spectra envelope in Fig. 11. Frequencies were in the range of 10 to

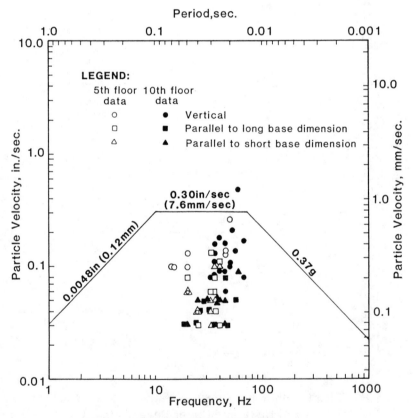

Fig. 11–Observed Building Response,
Building E

70 Hz. For all the recorded events on the fifth and tenth floors, the vertical component was much greater than the two horizontal components. A panel response (amplification) could contribute to this observed response.

The data from Building E is shown as a scaled distance plot in Fig. 12. Because of the absence of data from the basement of Building E, data is shown from other base level ground motion recorders in the general vicinity of Building E. Some of this base level data falls outside the

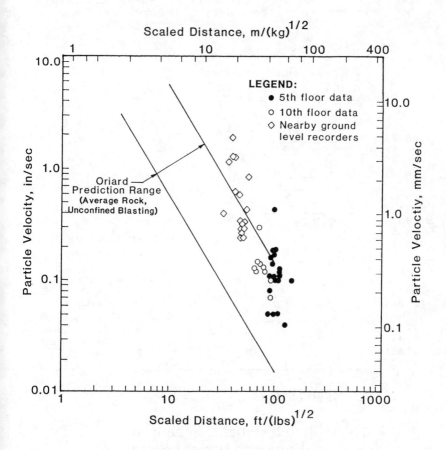

Fig. 12–Scaled Distance Comparison
of Base and Upper Level Data,
Building E

attenuation bounds found generally on the project. Possible
explanations include factors relating to site response,
response of the specific instrument location, or blasting
parameters, such as the estimate of the maximum charge
weight actually detonating at any given instant of time. As
before, a clearer representation of the relation between
base level and upper level data is shown as a bar chart in
Fig. 13. As in Building D, the range of peak particle
velocity at the base point below the recorders was estimated
as well as the most probable peak particle velocity at that
point. The most probable base amplitude vibration level
(directly below the recorder) is in all cases considerably
higher than the upper level data. The data from the two
upper levels is similar.

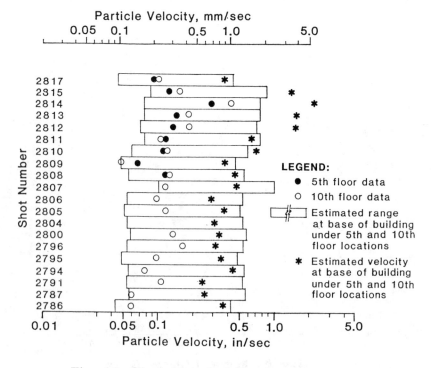

Fig. 13—Shot-by-shot Comparison
      of Base and Upper Level Data,
      Building E

DISCUSSION OF FINDINGS

A consistent finding during the monitoring program was that the vibrations input to the base of the structures were not amplified within the structure. There are three considerations which affect the response of structures near small scale blasting and which account for this finding:

1. The relatively small amount of available energy compared to other energy sources, such as earthquakes;
2. The distance from the energy source to the building, compared to the base dimensions of the structure;
3. The frequency of the ground vibrations relative to the natural frequency of the structure.

When all these factors are combined, the result is a significantly lower response compared to the highly amplified response that is sometimes feared by those who are not familiar with response of structures to nearby small-scale blasting. The individual considerations are discussed below.

Available Energy

In the particular case of tunnels near large structures, it is important to realize that the structure has great mass and the source has only a limited amount of energy. Only a few pounds of explosives were detonated on any given delay interval during blasting for the Peachtree Center Station. This quantity of explosives, and therefore, the energy released, was small in comparison to other energy sources, such as large-scale blasting operations or earthquakes.

Distance to Energy Source

Often engineers think of the base motion at a building foundation to be that of a rigid base, with a constant energy input along the base dimension. Such a model would be suitable in determining the response of a structure to an earthquake or other large energy source transmitting low-frequency vibrations at a distance of at least a few wave lengths. However, such assumptions are not valid for nearby, small scale tunnel blasting operations as indicated by base level recordings at more than one location for buildings B and C. For such sources, there is typically a significant attenuation of energy and change in wave form across the base itself. The degree of the attenuation is controlled by the distance from the nearest side of the foundation to the blasting and the base dimension of the foundation. The closer the energy source is to one side of the foundation, for any given level of particle motion, the more dramatic is the departure from the concept of a

rigid-base motion. Without an understanding of this
distinction between rigid-base motions and those generated
by small, nearby blasts, an understanding and a prediction
of high-rise building responses is impossible. One could
easily be misled by the numerical values of particle
velocities at the nearest portion of the building to the
blasting (the usual place for vibration monitoring). For
example, if our concern should be that for the total
structure, or even just the upper portions of a high-rise
structure, we would have less interest in a particle
velocity of 4 in/sec (102 mm/sec) generated by a blast at a
distance of 15 to 20 feet (5-7m) feet than we would have for
a particle velocity of 2 in/sec (51 mm/sec) generated by a
blast at a distance of 150 to 200 feet (50 - 70m).

Frequency of Vibrations

        The measured dominant frequencies of input base
motions were typical of those illustrated by Figs. 1 and 2.
These observed frequencies fall between approximately 10 Hz
and 120 Hz. The natural frequency of a structure commonly
refers to its motion in the horizontal plane. For the
structures monitored, natural frequencies would be estimated
as 2 Hz or less. The input frequencies measured are an
order of magnitude or more larger than the natural
frequencies of most high-rise structures. Conventional
spectra analysis techniques would indicate that no
amplification of whole-structure vibrations should occur in
this case. This prediction was essentially confirmed as
discussed in the case histories and may be summarized as
follows:
    1. Data for buildings A and B indicated that
        significant attenuation of vibrations occurred;
    2. Observed peak particle velocies for the fifth and
        tenth floors of Building E were less than estimated
        base level motions for all 20 cases;
    3. Observed peak particle velocies for the upper
        levels of Buildings C and D were less than those
        estimated for the base for the majority of cases
        (16 of 18 for Building C and 39 of 40 for Building
        D).

        Additional insight into building response may be
gained by considering relative attenuation of horizontal and
vertical components, the details of transducer placement,
and the possibility of individual panel response.
        Although there is no "natural frequency" defined for
vertical motion, consideration of the likely response to
vertical motions of various frequencies is also important.
Structural members must of necessity be rigid to support the
structure. Consequently, vibrations are transmitted
vertically through structural members essentially as rod
waves, with particle motion strongly influenced by the
material properties of the member. Therefore, in this case,

there can be relatively less vertical attenuation of vibration compared to the horizontal. Thus the fact that the vertical components of velocity are the largest for the majority of upper level responses for buildings C, D and E is not surprising. This condition may be contrasted with the base level vibrations in which the vertical component typically was the largest in approximately half the events recorded. Also, for building E, the fifth floor particle velocities exceed the tenth floor particle velocities in 7 of 9 and 8 of 9 cases for transverse and longitudinal directions, respectively. However, the tenth floor vertical component of particle velocity was greater than the fifth floor particle velocity in 7 of 9 cases.

Local panel amplifications in floors and walls can occur, and mask attenuation of vibration through the structure. Localized panel response and amplification seem to be indicated by the tenth versus fifth floor of Building E and the seventy-third versus fifteenth floor of Building C. As mentioned previously, sensors were mounted on floors to avoid defacing the structure. Panel response thus could have occurred and been recorded, effectively resulting in a conservative measure of structure response. Should an individual panel response be of interest, two considerations are important:

1. Attenuation of incoming vibrations affects the vibration input to individual panels;
2. Often vibrations imparted to the panels by small scale blasting are comparable to ambient vibrations within the building

Therefore, proper modeling of response of structures should consider the tendency for attenuation of vibrations through the structure.

Accelerations and Displacements

In examining the seismograph traces, the general wave form observed was typically more nearly triangular, rather than sinusoidal. Given the range of observed particle velocities and frequencies, the effect on a general estimate of likely accelerations and displacements was not critical. However, an individual estimate is significantly affected. For input velocities of 0.3 in/sec (7.62 mm/sec) - a commonly observed value - with frequencies in the range of 10 to 120 Hz, accelerations of approximately 0.05g to 0.6g would be estimated assuming sinusoidal vibrations and roughly 0.03g to 0.4g for the triangular wave assumption. Estimated ranges of displacements for the same input velocities would be 0.005 inch to 0.0004 inch and 0.004 inch to 0.0003 inch (0.10mm to 0.007mm) for sinusoidal and triangular waves, respectively. Thus, the assumption of sinusoidal wave forms is a conservative one, because it estimates values higher than the actual values.

Supplemental Findings

In addition to the data presented in the previous portions of the report, there were several other findings which became apparent as the monitoring work progressed. Initially, interest had been expressed in the water gel explosives used by the contractor on this project. Although well known to the industry, these explosives were new to the experiences of some of the persons associated with this project. Within the possible choices of common commercial explosive suitable for such a project, there is generally less than 10-15 percent variation in the effects that can be attributable to the explosive itself. This conclusion was confirmed for this project as well. Examination of data obtained on the project and comparisons generated by the water gel explosives were not unusual. The influence of this parameter is generally obscured by the more significant influence of parameters associated with the transmitting medium and the blast design.

Another matter of interest early in the project was the possibility of air overpressure (air blast) effects generating a response in the nearby structures. Ground vibrations travel through the ground with relatively high velocities, arriving at recording locations very soon after the blast. The air overpressures generated by a blast had to travel at a considerably lower velocity through the tunnel, up the nearest shaft, and then towards the building containing the recorder. The associated time delay with the arrival of the air overpressure would be detectable after the higher-frequency ground vibrations and include low-frequency components. In the cases discussed in this paper, structural response to air overpressure was not observed.

One other general aspect of the blasting and response of structures is worthy of mention, because of the urban setting of the project. This aspect is related to the human response to blasting effects. For small blasts detonated nearby, and where the rock is covered with a dense or relatively shallow overburden, the transmitted vibrations contain components that are well within the frequency range of human hearing. Thus, nearby listeners can hear the sound of these vibrations. Additional sound effects are generated within a structure due to its own, specific responses to the vibrations. Thus, an occupant of a nearby structure can hear the sound of the vibrations transmitted through rock and the structure, as well as additional responses, such as rattling of structural components or loose objects, windows, etc., (see Oriard (6)). For work in vented tunnels and shafts, the airborne sound could be heard following the sound generated by the building response to the ground vibration. For work in compressed air tunnels, no air-blast overpressure was transmitted outside the tunnel; hence, no airborne sound could be heard. Nevertheless, ground-transmitted sound was heard in nearby buildings.

CONCLUSIONS

The data obtained in this monitoring program provided the basis for several conclusions about the blasting on the MARTA project in downtown Atlanta and about the response of high-rise buildings to vibrations generated by nearby small-scale blasting:

1. Low frequency building responses were not observed, either from airblast or from ground vibrations for the multi-story structures which were monitored.

2. In the multi-story structures which were monitored, vibrations were generally attenuated in passing through the structures. Horizontal vibrations were attenuated significantly. Vertical vibrations were also attenuated, in some cases as significantly as horizontal vibrations.

3. Attenuation of vibrations across the base of the buildings was observed, indicating that it is not appropriate to use the concept of rigid-base input into large buildings to represent vibrations fromnearby construction blasting.

4. For high-frequency vibrations from nearby blasting, a first approximation of vibration intensity can usually be made for planning purposes using a scaled distance plot on the basis of the total distance from the blast to the location of interest within the building.

ACKNOWLEDGEMENTS

As mentioned previously, the vibration monitoring was for the Metropolitan Atlanta Rapid Transit Authority. Parsons Brinckerhoff/Tudor is the General Engineering Consultant for whom Law Engineering Testing Company is the Systemwide Geotechnical Consultant. The authors appreciate the cooperation of all parties involved during the course of the work, and appreciate their willingness to allow publication of this case study. The authors also acknowledge the cooperation of the Atlanta business community in allowing access to their buildings, and the efforts of their colleagues in Law Engineering in their gathering and analysis of data.

APPENDIX 1.-REFERENCES

Dowding, C. H. and Corser, P. G., "Cracking and
Construction Blasting: Importance of Frequency and
Free Response," Minimizing Deterimental Construction
Vibrations, ASCE National Convention Specialty Session,
April 14 - April 18, 1980.

Hendron, A. J. "Engineering of Rock Blasting on Civil
Projects," Structural and Geotechnical Mechanics,
Prentice-Hall, 1977.

Hudson, D. E., and Housner, G.W., "Structural
Vibrations Produced by Ground Motion," Transactions,
ASCE, 1955, and Proceedings Paper 816, October 1955.

Naik, T. R., "Predictions of Damage to Low-Rise
Buildings Due to Ground Vibrations Created by
Blastings," Vibrations of Concrete Structures, SP-60,
American Concrete Institute, Detroit, 1979,
pp. 249 - 264.

Oriard, L. L., "Blasting Operations in the Urban
Environment," Bulletin of the Association of
Engineering Geologists, Vol. IX, No. 1, Winter, 1972,
pp. 27 - 46.

Oriard L. L., "Blasting Effects and Their Control,"
prepared for "Handbook on Underground Mining Methods,"
sponsored by SME of AIME, 1980.

ENERGY - ATTENUATION RELATIONSHIPS FROM CONSTRUCTION VIBRATIONS

Richard D. Woods[1], M.ASCE
and
Larry P. Jedele[2], M.ASCE

ABSTRACT

Vibrations generated by various construction operations are often detrimental to the surrounding neighborhood. Data has been accumulated from actual construction projects and proposed construction sites. A relationship has been derived between impact energy (or source energy) and level of vibration for specific distances from the source in soils of various classifications. Attenuation coefficients have been determined experimentally at many construction sites and these coefficients used to propose a basic earth materials classification system.

Cases studied include dropping heavy weights in industrial operations like breaking up scrap cast iron and construction activities like dropping weights for dynamic compaction and pile driving. Also, the vibrations caused by highway construction equipment, finished highway traffic, and railroad traffic on a very sensitive manufacturing operation are included.

The data presented basically confirms trends presented by others and reinforces the "scaled-distance" approach for construction generated vibrations.

INTRODUCTION

Vibrations generated by various construction operations are often detrimental to the surrounding neighborhood for several reasons. At one end of the spectrum, some vibrations are severe enough to do physical damage to structures, while at the other end of the spectrum, some very small vibrations are sufficient to impede the operation of sensitive instruments or are disturbing to people. Considerable data has been collected and published with respect to vibrations from blasting operations, for example Oriard (1972), Hendron and Oriard (1972), Dowding (1971) and others. Some data has been collected and published for construction vibrations, principally Wiss (1968, 1974 and 1981). Carefully monitored operations can still be useful in extending knowledge in this area. In particular, it is still difficult to find data relating input energy to amplitudes of vibrations at specific distances.

---

1 Professor of Civil Engineering, Univ. of Michigan, Ann Arbor, Michigan and Principal, Stoll, Evans, Woods and Assoc., Ann Arbor, Michigan.

2 Staff Engineer, Stoll, Evans, Woods and Associates, 111 W. Kingsley, Ann Arbor, Michigan 48103.

Two categories of problems can be identified with respect to construction vibrations: prediction of the rate of decay (or attenuation) of vibrations from a location of known vibration amplitude and prediction of the amplitude at a specific distance due to a known energy source. In this paper we will present new data from construction sites where vibrations have been measured, and attenuation coefficients determined for sources of known energy input. This data is compared with previously published data and presented in an energy-distance-amplitude format which is very useful to the engineer faced with predicting the vibrations which might result from some construction operation or activity.

VIBRATION ATTENUATION DATA

Some well known ways of presenting vibration data in which the reduction of amplitude with distance (attenuation) is included are presented first. These expressions, however, do not include the energy level of the vibration source.

The decay of amplitude of vibrations with distance can be attributed to two components, geometric damping (radiation damping) and material damping (hysteretic damping) as in Richart, Hall and Woods (1970) for example. In the simplest form geometric damping can be described by the equation:

$$w_2 = w_1 \ (r_1/r_2)^n \qquad (1)$$

in which,

w_1 = amplitude of vibration at distance from the source, r_1 .
w_2 = amplitude of vibration at distance from the source, r_2 .
r_1 = distance to point of known amplitude.
r_2 = distance to point of unknown amplitude.

n = coefficient depending on type of wave propagated, where
  $n = 1$ for body waves in the ground,
  $n = 2$ for body waves along surface,
  $n = 1/2$ for Rayleigh waves.

If material damping is included, another term is added to equation (1) making it:

$$w_2 = w_1 \ (r_1/r_2)^n \ \exp \ [- \ a(r_2-r_1)] \qquad (2)$$

in which,

r_1, r_2, w_1, w_2, and n as in equation (1),
$\exp$ = base of natural logarithm (e), and
a = coefficient of attenuation (units of 1/distance).

Figure 1 illustrates the relationship between actual vibrations measured in the field and the expressions of equations 1 and 2. This

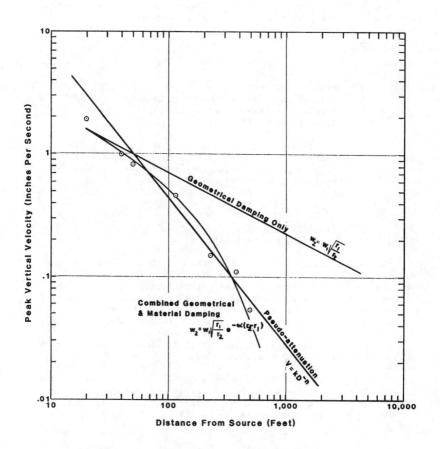

FIG. 1 – Various Forms for Vibration Attenuation

log-log plot of peak vertical particle velocity vs distance from the
source of vibration illustrates that the curved line for geometrical
and material damping provides a better representation of decay of
vibration with distance than a straight line. However, for close-in
distances, the straight line approximation may be adequate.

Wiss (1981) suggested a 'best fit' straight line on a log-log
amplitude-distance plot with the following formula:

$$V = k\ D^{-n} \tag{3}$$

in which,

V = peak particle velocity,
k = value of velocity at 1 unit of distance,
D = distance from the vibration source,
n = slope or attenuation rate.

This 'n' rate is not 'classical' attenuation but may be considered a
pseudo-attenuation coefficient. The data from the case studies
associated with this paper have been presented in both 'α' and 'n'
forms giving a comparison which may be instructive, see Figs. 2 to
10.

PROPOSED CLASSIFICATION OF EARTH
MATERIALS BY ATTENUATION COEFFICIENTS

The senior writer has collected data on attenuation coefficients
(α) in the format of equation (2) for about 20 years, and has
combined his data and other published data linking soil type with
attenuation coefficient into Table 1. Data from 36 sites in materials
ranging from sound, hard rock to very soft clay and loose sand are
included in Table 1. Four classes of materials have been identified
in Table 1 and brief descriptions given of materials belonging in
each class. These classes are subjective and some other choices might
be made if additional data were included.

Since the coefficient α in EQ 2 is frequency dependent (Richart
et al, 1970), values are listed in Table 1 for two frequencies, 5 Hz
and 50 Hz. For other frequencies, α can be computed very simply
from:

$$\alpha_2 = \alpha_1\ (f_2/f_1) \tag{4}$$

where,

α_1 = known value of α at frequency f_1, and
α_2 = unknown value of α at frequency f_2.

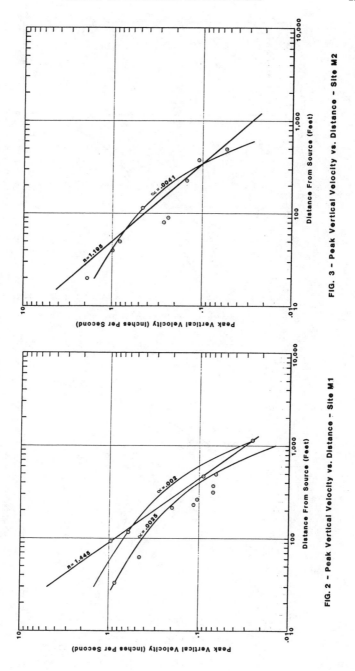

FIG. 3 – Peak Vertical Velocity vs. Distance – Site M2

FIG. 2 – Peak Vertical Velocity vs. Distance – Site M1

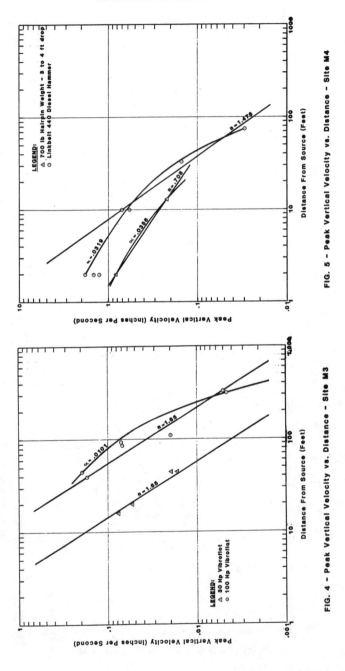

FIG. 5 – Peak Vertical Velocity vs. Distance – Site M4

FIG. 4 – Peak Vertical Velocity vs. Distance – Site M3

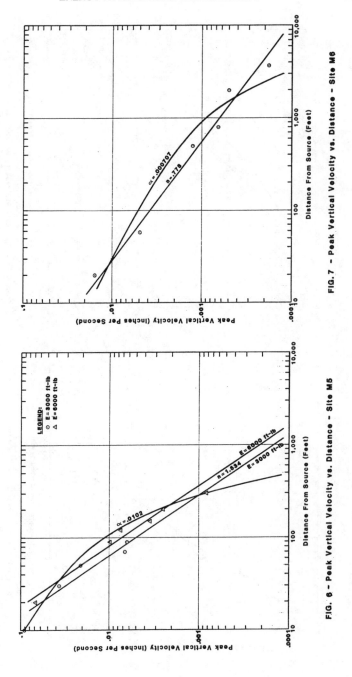

FIG. 7 – Peak Vertical Velocity vs. Distance – Site M6

FIG. 6 – Peak Vertical Velocity vs. Distance – Site M5

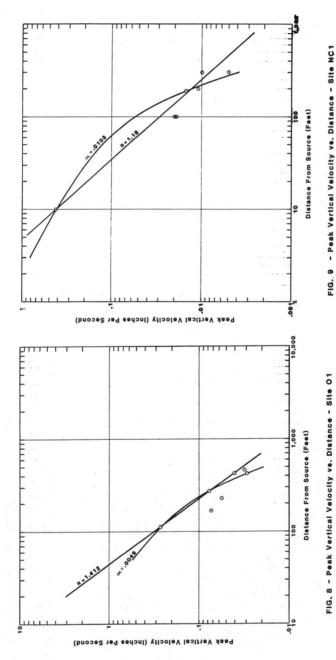

FIG. 9 – Peak Vertical Velocity vs. Distance – Site NC 1

FIG. 8 – Peak Vertical Velocity vs. Distance – Site O 1

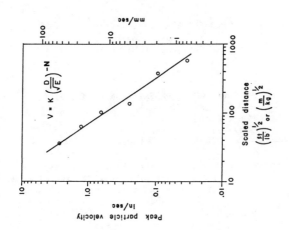

$$V = K \left(\frac{D}{\sqrt{E}}\right)^{-N}$$

Scaled distance
$\left(\frac{ft}{lb}\right)^{1/2}$ or $\left(\frac{m}{kg}\right)^{1/2}$

Peak particle velocity
in/sec

mm/sec

**FIG.11 - Typical Form: Peak Particle Velocity vs. Scaled Distance
From Ref 8**

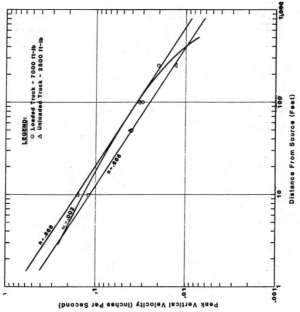

LEGEND:
O Loaded Truck - 7000 ft-lb
△ Unloaded Truck - 2500 ft-lb

Peak Vertical Velocity (Inches Per Second)

Distance From Source (Feet)

FIG. 10 - Peak Vertical Velocity vs. Distance - Site NC2

TABLE 1

PROPOSED CLASSIFICATION OF
EARTH MATERIALS BY
ATTENUATION COEFFICIENT

CLASS	ATTENUATION COEFFICIENT α (1/ft) 5 Hz 50Hz		DESCRIPTION OF MATERIAL
I	0.003	0.03	Weak or Soft Soils-lossy soils, dry or partially saturated peat and
	to	to	muck, mud, loose beach sand, and dune sand, recently plowed ground, soft
	0.01	0.10	spongy forest or jungle floor, organic soils, toposoil.
			(shovel penetrates easily)
II	0.001	0.01	Competent Soils- most sands, sandy
	to	to	clays, silty clays, gravel, silts,
	0.003	0.03	weathered rock. (can dig with shovel)
III	0.0001	0.001	Hard Soils- dense compacted sand, dry
	to	to	consolidated clay, consolidated glacial
	0.001	0.01	till, some exposed rock. (cannot dig with shovel, must use pick to break up)
IV	<0.0001	<0.001	Hard, Competent Rock- bedrock, freshly exposed hard rock.
			(difficult to break with hammer)

PARTICLE VELOCITY RELATED TO SOURCE ENERGY

Wiss (1981) has suggested a simple expression for impact energy sources relating energy and particle velocity as follows:

$$V = C \ (E)^{\beta} \qquad (5)$$

in which

V = peak particle velocity,
C = velocity at energy of 1 unit,
E = impact energy, and
β = slope of velocity increase (given in Wiss 1981 as α).

This expression is appropriate for a given distance from the source. Wiss suggests that energy and distance could be coupled in a "scaled-distance" format similar to that used for explosives. The equation is as follows:

$$V = k \left[\frac{D}{\sqrt{E}} \right]^{-N} \qquad (6)$$

in which,

V = peak particle velocity
k = intercept at 1 energy unit
D = distance from source
E = energy of source
N = slope (given in Wiss 1981 as n).

Figure 11 from Wiss (1981) shows an example of a plot of equation (6).

PROJECTS INCLUDED IN SOURCE-ENERGY STUDY

A brief description of the construction related projects from which amplitude-distance-energy data were collected are given below. Items presented include the general purpose of the vibration investigation, soil type(s), energy source and magnitude if known, and other related details. The projects are identified according to the State in which they are located and a number to account for more than one in a state. Table 2 summarizes the description of each site and the data obtained at that site.

In each study vibrations were measured with velocity transducers (calibrated geophones) with natural frequencies of one Hz or four to eight Hz. The vibration records were obtained on either a dual trace, storage screen oscilloscope or a dual channel, hot-pen writing, oscillographic strip chart recorder. The accuracy of the data obtained in this manner is thought to be within +/- 25%.

(1) **Michigan #1** (M1). Construction of one and two story retail stores and adjacent parking areas was proposed in an area with as much as 40 feet (12.2 m) of loose foundry sand and miscellaneous fill. Dynamic compaction was selected to densify the loose sands in order to provide support for these structures. For this purpose, a 15 ton (133 kN) weight was dropped 80 feet (24.4 m) at pre-determined locations until the desired degree of compaction was achieved. The attenuation of vibration across the site and the effects of vibration on nearby residential structures were investigated.

(2) **Michigan #2** (M2). The relocation of an existing metal reclamation operation required an investigation for attenuation of vibrations generated from this process. The investigation was performed because a high pressure gas main and utility tunnel were located in the proximity of the new metal reclamation facility. These utilities could be affected by the resulting vibrations created from the impact of an 18-ton (160 kN) ball dropping 60 ft (18.3 m) which was planned for the process at the new facility. For this study attenuation of vibrations was determined by measuring the vibrations from impact of a 6 ton (53 kN) ball dropping onto the natural ground from various heights ranging up to 30 feet (9.1 m). The soil conditions consisted of about 100 feet (30 m) of soft till clay overlying limestone bedrock.

(3) **Michigan #3** (M3). Densification of loose sand by vibroflotation was performed to provide needed support for an addition to an existing structure. However, if the vibroflotation probe was in close proximity to the existing structure, the resulting vibrations could adversely affect the structure. Therefore, the purpose of the investigation was to plan the vibroflotation activity based on vibration attenuation criteria developed from both 30 Hp and 100 Hp vibroflots.

(4) **Michigan #4** (M4). A vibration study on the effects of driving sheetpiling in an area consisting of about 20 ft (6.1 m) of soil overlying bedrock was conducted. Pile driving hammers included a 700 lb (3.1 kN) hairpin shaped weight dropped several feet onto the sheets and a Linkbelt 440 diesel hammer. Vibrations created by these hammers were generally measured at the ground surface. However, some vibrations were measured directly on the exposed rock while driving sheeting into the rock. Vibrations from the diesel hammer were measured at varying sheeting penetrations into the soil overburden in addition to driving into bedrock.

(5) **Michigan #5** (M5). The construction of a plant to contain large blanking and forming presses was being planned. In light of the vibrations that would be created from these presses, a study was done to determine if any troublesome (with respect to human perception) vibrations would extend beyond the owner's property limits. Vibration attenuation data was compiled on the plant site by dropping a 1.5 ton (13.3 kN) ball from heights up to two feet (.61 m) on a 4.3 feet diameter (1.3 m) concrete plug cast at the bottom of a 13 feet (4.0 m) deep and 6 ft (1.8 m) diameter shaft in stiff to hard clays.

(6) __Michigan #6__ (M6). Vibrations were monitored from a combined pump-generator unit at a pumped storage facility. The study was conducted to determine the effects, if any, that current and future pumping operations would have on the stability of deep sands and till clays which formed a bluff on which private homes were located.

(7) __Ohio #1__ (O1). Vibration levels were determined from an existing metal reclamation facility. Metal was crushed on the site with an 8 ton (71.2 kN) ball dropped from varying heights up to 30 ft (9.2 m) onto an anvil foundation consisting of steel dies and crushed rock. The soils in the area were reported to be lacustrine clays. The vibrations were measured to assess their effect on existing structures including a truck weighing scale.

(8) __North Carolina #1__ (NC 1). At a site consisting of deep loose, fine sands, the attenuation coefficient was determined to evaluate the influence of railroad and highway traffic on a sensitive manufacturing operation. Ground vibrations were generated by dropping a 300 lb (1.3 kN) casing driver from a height of 5 ft (1.5 m) on the ground surface and measuring vibration amplitudes at four distances out to 300 feet (91.5 m).

(9) __North Carolina #2__ (NC 2). To evaluate the effects of highway traffic on a sensitive manufacturing process, a dump truck, both loaded and empty, was driven over a plank on an asphalt pavment and vibrations were measured at varing distances from the edge of the road. The soil at this site was loose, fine sand for considerable depth with a water table at about 6 ft(1.8 m) below ground surface.

VELOCITY-DISTANCE-ENERGY RELATIONSHIPS

The peak particle velocity-distance data for the sites described in Table 2 and Figs. 2-10 are presented in Fig. 12. Coefficients 'n' in EQ 3 have been determined from the 'best-fit' curve for each site and three distinct ranges of slope were identified , namely n = 1.45-1.53, 1.15-1.20, and 0.66-0.78. Using EQ 5 relating velocity and energy, β factors can be determined for the data with a common range of n. The ranges for β parallel those for n and are β =0.746-0.816 corresponding to n = 1.45-1.55 and β =0.474-0.513 for n = 1.15-1.20. Correlating a range in β for the n=0.66-0.78 grouping was not possible because data available at this time did not cover a wide enough range in energy levels. It appears that the ranges in 'n' and 'β' are independent of soil type, energy source, and energy level.

However, a more useful format for the velocity-distance- energy relationship is that provided by the 'scaled-distance' approach. The data used to construct Fig. 13 is also used to construct Fig. 14. Wiss (1981) suggested that the slope, N (see EQ 6), of the line on Fig. 14 ranges between 1.0 and 2.0 with an average of about 1.5. Most of the data in this study shows a slope, N = 1.52 for n = 1.41 to 1.53. Some of the data reported in this study, however, differs from the usual, and on Fig. 14, for the n range of 1.15-1.20, an 'N' of 1.11 is found.

TABLE 2: VIBRATION - ATTENUATION DATA FROM EACH SITE

SITE	SOIL TYPE	ENERGY SOURCE/ DESCRIPTION	ENERGY OR POWER	FREQUENCY RANGE (Hz)	ATTENUATION FACTOR α(1/ft)	* ADJUSTED TO f = 5 Hz α	PSEUDO-ATTENUATION FACTOR - n
M1	Sand	Drop Weight/ 15 tons-80' drop	2,400,000 Ft-lb	5-10.5	.002-.0025	.0022	1.445
M2	Clay	Drop Weight/ 6 tons-30' drop	360,000 Ft-lb	8.5-17	.0041	.0016	1.195
M3	Sand	Vibroflot	100 Hp 30 Hp	31 31	.0101	.0016	1.65 1.65
M4	About 20' soil over bedrock	Diesel Pile Hammer 700 lb. Hairpin	18,200 Ft-lb 2,500 Ft-lb	18-33 40-44	.0319 .0356	.0063 .0043	1.52 0.698
M5	Clay	Drop Weight 13' below G.S./1.5 ton- 1' & 2' drop	6,000 Ft-lb 3,000 Ft-lb	12-33 30-48	.0102	.0023	1.476
M6	Sand	Pump-Generator	-	11-60	.000707	.0001	0.778
O1	Clay	Drop Weight/ 8 ton-30' drop	480,000 Ft-lb	9-12	.0049	.0023	1.412
NC1	Sand	Drop Weight/ 300 lb-5' drop	1,500 Ft-lb	20-40	.0103	.0017	1.15
NC2	Sand	Dump Truck driving over 3" high plank	7,000 Ft-lb(loaded) 2,500 Ft-lb(unloaded)	10-40	.002	.0004	0.666
Ref 2	-	1 lb. of Dynamite	980,000-2,050,000** Ft-lb	-	-	-	1.483

* See Equation (4)
** Estimated on basis of information on pg. 70 of Blaster's Handbook (Ref 2)

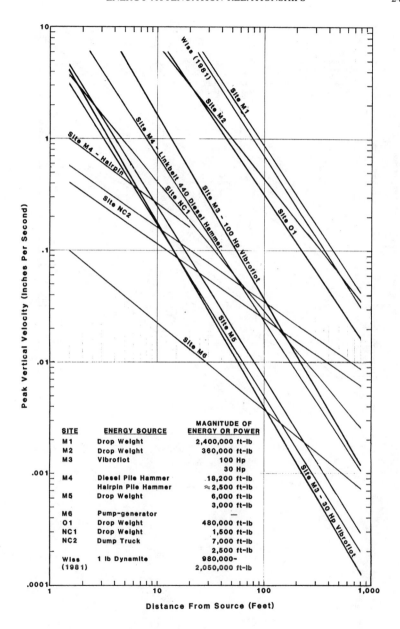

SITE	ENERGY SOURCE	MAGNITUDE OF ENERGY OR POWER
M1	Drop Weight	2,400,000 ft-lb
M2	Drop Weight	360,000 ft-lb
M3	Vibroflot	100 Hp
		30 Hp
M4	Diesel Pile Hammer	18,200 ft-lb
	Hairpin Pile Hammer	≈ 2,500 ft-lb
M5	Drop Weight	6,000 ft-lb
		3,000 ft-lb
M6	Pump-generator	—
O1	Drop Weight	480,000 ft-lb
NC1	Drop Weight	1,500 ft-lb
NC2	Dump Truck	7,000 ft-lb
		2,500 ft-lb
Wiss (1981)	1 lb Dynamite	980,000- 2,050,000 ft-lb

FIG. 12 - Magnitude of Construction Related Vibrations

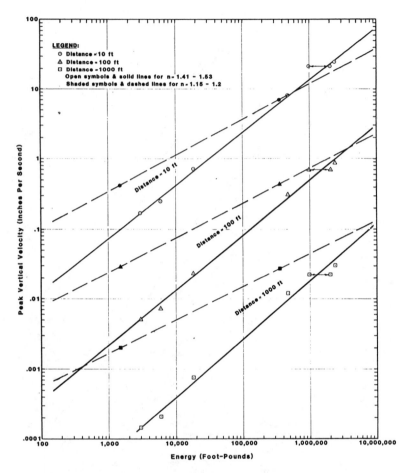

FIG. 13 - Peak Vertical Velocity vs. Energy

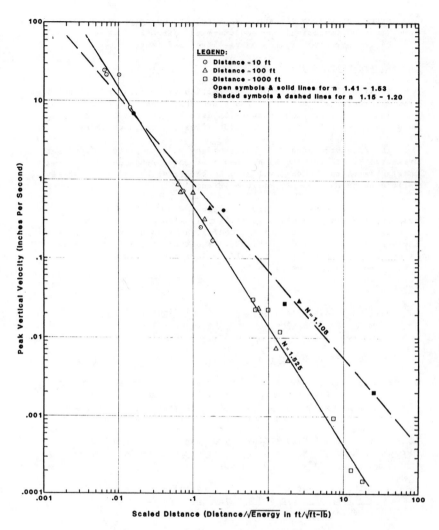

FIG. 14 – Peak Vertical Velocity vs. Scaled Distance

From the 'scaled distance' approach, the peak particle velocity (amplitude of vibration) can be determined directly from a plot like Fig. 14 for any given source energy level and range.

SUMMARY AND CONCLUSIONS

Additional data relating construction vibrations to distance from the source and energy of the source has been presented. Also, a compilation of vibration attenuation data has been presented and suggestions made for a material classification based on damping.

The data presented does not show any major departure from that presented by other authors, but confirms most trends and substantiates the general validity of the velocity–distance– energy relationship in terms of 'scaled–distance'.

Although the data confirms recognized approaches to data presentation and prediction equations, it also presents some situations in which conventional coefficients do not provide a satisfactory correlation. This simply points to the fact that data needs to be collected in the field to determine whether or not a given site fits the general pattern.

REFERENCES

1. Dowding, C.H. (1971), Response of Buildings to Ground Vibrations Resulting from Construction Blasting., Ph.D. Thesis, University of Illinois, Urbana, Ill.

2. Dupont (1977), Blaster's Handbook.

3. Hendron, A.J., Jr. and Oriard, L.L. (1972), "Specifications for Controlled Blasting in Civil Engineering Projects," Proceedings, North American Rapid Excavation and Tunneling Conference, Chicago, June 5-7, Vol. 2, pp. 1585-1609.

4. Oriard, L.L. (1972), "Blasting Operations in the Urban Environment," Bulletin of the Association of Engineering Geologists, Vol. IX, No. 1, pp. 27-46.

5. Richart, F.E., Jr., Hall, J.R., Jr. and Woods, R.D. (1970), Vibrations of Soils and Foundations, Prentice-Hall, Englewood Cliffs, N.J., 414 pp.

6. Wiss, J.F. (1968), "Effects of Blasting Vibration on Buildings and People," Civil Engineering, ASCE, July, pp. 46-48

7. Wiss, J.F. (1974), "Vibrations During Construction Operations," Journal of the Construction Division, ASCE, Vol. 100, CO3, Sept., pp. 239-246.

8. Wiss, J.F. (1981), "Construction Vibrations: State-of-the-Art," Journal of the Geotechnical Engineering Division, ASCE, Vol. 107, GT 2, Feb., pp. 167-181.

GROUND VIBRATIONS DURING DYNAMIC COMPACTION

by Paul W. Mayne[1]

ABSTRACT

Ground vibration data taken at twelve different dynamic compaction sites are reviewed for comparative purposes. Weight sizes ranged from 3 to 45 tons and drop heights varied from 5 to 100 feet. Soil types were generally granular materials (silty sands to rockfill). Scaled distance graphs based on the square root of the applied energy per blow appear applicable, yet possibly conservative for the high energy systems reviewed in this study. The sites also show very similar attenuation of particle velocities normalized to the impact velocity versus distance normalized by the radius of the weight. Since the observed frequencies of vibration are in the low range of 2 to 20 hertz, an important consideration in the measurement of vibrations due to dynamic compaction is that many commercial seismographs have transducers with nonlinear responses below 6 hertz. In addition, a lower threshold velocity than the commonly accepted 2 ips is warranted since recent studies have shown that low frequency transient vibrations are potentially more damaging than high frequency vibrations.

INTRODUCTION

During the last several years, dynamic compaction has become popular as an effective method of improving loose sands and granular fills insitu (4, 10, 11, 12, 15, 16). The procedure involves systematically dropping a large steel or concrete weight onto the ground surface to densify the underlying soils. One undesirable side effect is the generation of ground vibrations which emanate from the point of impact (5, 12, 15). Since dynamic compaction is an attractive economical solution, its use is seen increasingly in urban and suburban communities where real estate costs are high.

Ground vibrations can be potentially damaging to nearby building structures and sensitive equipment, as well as annoying to people. Consequently, careful and proper monitoring of ground vibration levels and vibration frequencies must be made in order to protect all interested parties. Ground vibrations caused from dynamic compaction operations are unique from other types of construction activity, such as blasting, pile driving, and traffic. In this regard, vibrations from dynamic compaction are characterized by low-frequency waves which are (1) potentially more damaging than high-frequency waves and (2)

---

1. Senior Geotechnical Engineer, Law Engineering Testing Company, Post Office Drawer QQ, McLean, VA 22101

below the frequency range of many commercially available vibration
monitor seismographs.

The purpose of this paper is to discuss the field measurement and
analysis of ground vibrations during dynamic compaction. This
includes a review of seismograph equipment, measurement limitations,
threshold vibration criteria, and a summary of ground vibration data
which were previously obtained at 12 different dynamic compaction
sites.

VIBRATION MEASUREMENT

The magnitude of ground vibration levels may be measured in terms of
displacement(s), velocity (v), or acceleration (a). If the time
history of the waveform is known, then numerical or digital integration and differentiation may be used to relate s, v, and a as
functions of time t:

$$a = \frac{dv}{dt} = \frac{d^2s}{dt^2} \tag{1}$$

Often, for simplicity sake, harmonic motion is assumed in converting
from one mode to another. Real motions are almost always more
complex, irregular, and variable than simpler sinusoidal waveforms.
However, since one often deals with orders of magnitude and logarithmic scales in vibration measurement, an approximate analysis may in
fact be sufficient for many purposes.

The relationships among peak values of harmonic waves may be expressed
by:

$$a = 2\pi f v = (2\pi f)^2 s \tag{2}$$

where f = frequency of vibration.

For most construction-related vibrations, the velocity at a point on
the ground (the particle velocity) has been shown to be the best
indicator of damage potential and annoyance levels (2, 6, 13, 19, 21,
22, 25). For certain situations, a combination of velocity and
displacement measurements may be appropriate. Possibly, the choice of
particle velocity is related to the observed frequency range of
transient vibrations occurring in construction, which are typically
between 5 and 200 Hz. For comparison, seismologists studying earthquakes, which have frequencies of about 1 to 2 Hz or less, use
accelerometers. Also, for alignment and performance monitoring,
mechanical engineers often use spectrum analyzers to measure dynamic
displacement levels of machines (typically f > 100 Hz).

Routine field measurements are taken using a vibration monitor seismograph. Usually, the seismograph package includes a triaxial component
transducer or geophone, electronic signal conditioners, and a recording mechanism. A review of several leading commercial units which are
available has been prepared by Stagg and Engler (21). Most commonly,
the ground vibration records are written on oscillographic paper or

magnetic tape, although a few units provide an electronic digital
display output or ticker-tape summary of the vibration levels. Most
units record the complete waveform of the measured vibration. Often,
the wave is believed to be of the Rayleigh-wave type, although compression, shear, and Love waves also exist. One recent seismograph
unit on the market also has a built-in microprocessor to provide fast-
Fourier transforms and spectral analysis. Data recorded on magnetic
tape also allows a spectrum analysis. However, routine field measurements have not yet developed to this stage and a discussion of
frequency spectral methods are beyond the scope of this study. The
seismograph unit currently used by the author is reportedly accurate
to within + 10 % for vibration amplitudes at 30 hz. Timing marks are
claimed accurate within 3 %.

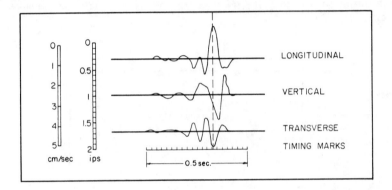

Fig. 1   Example trace of vibration record from dynamic
compaction in Morris, Alabama (W = 20.9 tonne,
H = 18.3 m, distance = 12.2 m).

Vibration measurements are taken in three mutually orthogonal directions simultaneously (vertical, longitudinal, and transverse axes).
An example vibration recording taken during dynamic compaction is
presented in Fig. 1. Often, the peak value of each directional
component is sought. Beyond this, unfortunately, data are presented
in a variety of ways by different individuals. Damage criteria
developed by the Bureau of Mines (19) for blasting have been based
upon the maximum single value of the three directional components
(x_{max}, y_{max} or z_{max}). Since real waves are three-dimensional
and the transducer axes may not be exactly in line with the source of
vibrations, some engineers (21, 22) prefer to calculate the true
vector sum (TVS) of the triaxial components:

$$TVS = \sqrt{(x_t)^2 + (y_t)^2 + (z_t)^2} \qquad (3)$$

where all values are obtained at the same time t. Some seismograph equipment presents the vibration data directly in the TVS format. Mistakenly, several individuals (3, 20) have expressed the vibration levels in terms of the pseudo vector sum (PVS):

$$PVS = \sqrt{x_{max}^2 + y_{max}^2 + z_{max}^2} \qquad (4)$$

It is noted, however, that x_{max}, y_{max}, and z_{max} rarely, if ever, occur at the same time. At most, the PVS could be 73% higher than the maximum single component velocity. For the example vibration in Fig. 1, the peak single component, TVS, and PVS are 0.62, 0.69, and 0.83 ips, respectively (16,17, and 21 mm/sec). Typically, the TVS values are about 10 to 40% higher than the maximum single component velocity, as shown by Fig. 2.

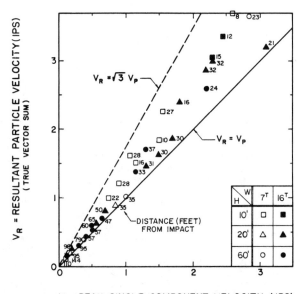

Fig. 2   Observed trend between measured true vector sum velocity and peak single component velocity from Tampa, Florida.

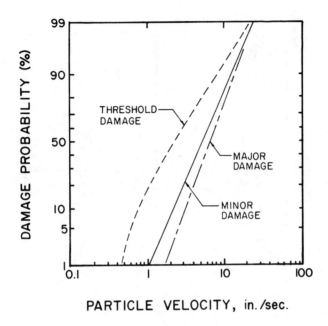

PARTICLE VELOCITY, in./sec.

Fig. 3   Probability analysis of damage potential from transient
         ground vibrations (data from surface mine blasting, U.S.
         Bureau of Mines, 1980, ref. 19).

DAMAGE CRITERIA

For many years, a limiting peak particle velocity of two inches per
second (50 mm/sec) has been considered the structural damage criteria
for one and two-story buildings. The primary sources of data for this
basis came from blasting records from surface mining operations near
residential communities. Higher and lower limits were proposed for
larger structures and older sensitive structures, respectively (2).
Despite the use of a 2 ips criterion, numerous litigation claims and
complaints were filed in the courts. Consequently, a re-evaluation
study of vibration damage was performed by the Bureau of Mines and
published in 1980 (19). Previous data and new data were analyzed
using three different methodologies: (1) statistical mean and
variance, (2) probability theory, and (3) observational. Damage was
classified according to three types: threshold, minor, and major
categories, as indicated by the probability of damage graph shown in
Fig. 3. The extensive review culminated in a combined particle
velocity-displacement criterion, shown as Fig. 4.

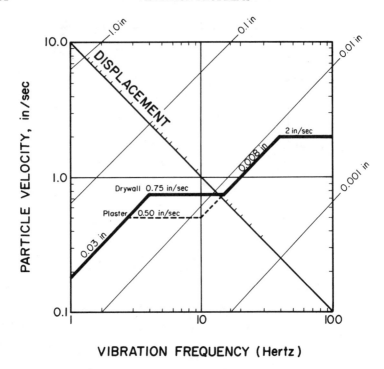

VIBRATION FREQUENCY (Hertz)

Fig. 4   Combination velocity-displacement criterion (U.S. Bureau of Mines, 1980, ref. 19).

Several commercial seismographs are capable of measuring either displacement, velocity, or acceleration, although not simultaneously. Alternatively, displacements could be estimated using Eq (2), especially since the axes in Fig. 4 assume harmonic motion. The use of the damage criterion in Fig. 4 for dynamic compaction operations could be questioned since it was developed for blast induced vibrations. However, a separate and independent study of vibration damage from blasting, pile driving, and machine sources resulted in similar criteria where limiting particle velocities depend upon frequency (22).

The significant point relevant to dynamic compaction is that low-frequency transient vibrations are potentially more damaging to structures than higher frequency vibrations. Most construction vibrations, blasting operations, and pile driving cause vibrations with frequencies between 5 and 200 Hz. For dynamic compaction, however, Mitchell (16) has indicated that frequencies are typically between 2 and 20 Hz.

FREQUENCY CONSIDERATIONS

Ideally, the vibration frequencies of a waveform should be determined from spectral analysis on Fourier transforms. Practically, however, most equipment available today does not provide this information, especially for field work requiring immediate decisions. Vibration frequencies may be approximately determined by scaling the individual periods from the waveform (21) or by an averaging method by counting the major peaks within a specified time duration of the waveform (22).

Histograms of vibration frequencies obtained by the author at two sites are presented in Fig. 5. The Tampa site was underlain by loose sands with a high groundwater table. For distances between 6 and 57 feet, the mean vibration frequency was 7.5 Hz with standard deviation of + 4.2. At Birmingham, dynamic compaction was used to densify coal spoil material with groundwater over 100 feet deep and D_{50} = approximately 2 inches. The mean and standard deviation of observed frequencies were 10.5 and 2.8 Hz, respectively. Pearce (17) and Leonards et al (10) have also indicated typical vibration frequencies of 5 to 8 Hz for dynamic compaction operations.

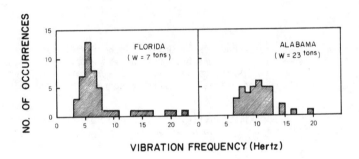

Fig. 5   Histograms of vibration frequencies from dynamic compaction operations at two sites.

Low-frequency vibrations present another problem for those responsible for monitoring them. Many commercial seismographs cannot directly measure vibration levels when the frequency of vibration is less than 5 or 6 Hz. The restriction is primarily due to the resonant frequency and damping characteristics of the transducer. Several manufacturers provide a magnification factor for determining the vibration amplitude when the vibration frequency falls below the specified frequency range of the equipment. The gain factors of two commercial units shown in Fig. 6 indicate that the measured particle velocities may be wrong by a factor of 5 or more unless the vibration frequency is known. The author knows of at least one seismograph unit available which does not measure vibration frequency at all! Readers are cautioned to check the manufacturer's specifications regarding the applicable range of the equipment used.

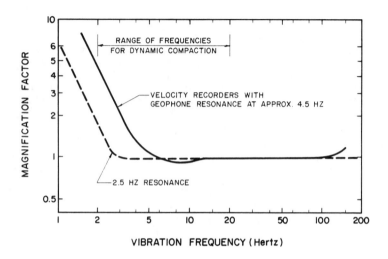

Fig. 6   Frequency dependence of magnification factor for
          commercial seismographs. Transducer resonance may result
          in unconservative measurement unless vibration frequency
          is known and vibration amplitude corrected accordingly.

As a first order approximation (14, 18) the vibration frequency (f_n)
from dynamic compaction operations may be estimated as:

$$f_n = \frac{1}{T} = \frac{1}{2\pi}\sqrt{\frac{k}{m}} \qquad (5)$$

where   T = period of vibration

$$k = \frac{4\,G\,r_0}{1-\nu} = \text{vertical stiffness of the system (18)}$$

         G = shear modulus

         r_0 = radius of the mass

         ν = Poisson's ratio

         m = mass of weight = W/g

         g = gravitational constant = 32 ft/sec^2 = 9.8 m/sec^2

Eq (5) indicates that low frequency vibrations are associated with loose soils (with low shear moduli) and for larger weights. The deceleration-time histories of two impacts during dynamic compaction are shown in Fig. 7. The decelerations were measured by mounting an accelerometer at the center and top of a 23-ton steel/concrete weight (14). The accelerometer output was transmitted to an oscilloscope by cable and the image recorded on polaroid film. With groundwater at considerable depth, each successive blow of the weight densified the sandy gravelly soils. The observed half-period is approximately 50 msec or, T = 0.1 sec, indicating a frequency of vibration of about 10 hertz. This is consistent with the observed mean frequency from particle velocity monitoring previously presented in Fig. 5 and taken at distances of 40 to 350 feet from impact.

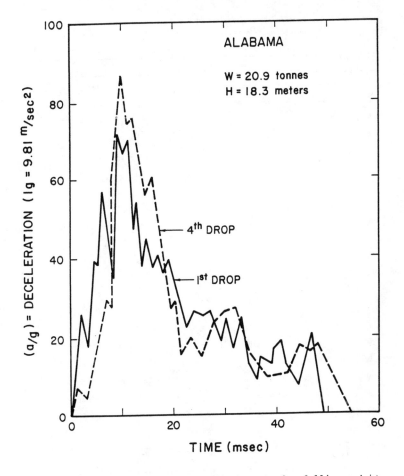

Fig. 7   Accelerometer measurement during impact of a falling weight.

For sites with a high groundwater table, localized liquefaction may occur around the point of impact. Consequently, excess pore pressures develop, effectively reducing the soil stiffness and causing low-frequency vibrations. This may explain the observed low-frequencies at Tampa (see Figure 5). The potential for damage increases at such a site since the level of shear strain may be high. Shear strain amplitudes may be measured in the field as the ratio of peak particle velocity (PPV) to shear wave velocity (Vs) of the soil medium. Unfortunately, since shear wave velocity and shear modulus are related, Vs also decreases with higher levels of shear strain.

TABLE I.

DYNAMIC COMPACTION SITES WITH GROUND VIBRATION DATA

Site Location	Soil Type	Weight in Tons (tonnes)	Drop Height in Feet (m)	Reference Source
◇ Birmingham, Alabama	coal spoil	23 (20.9)	60 (18.3)	This study
△ Alexandria, Virginia	clayey sand and gravel fill	7.8 (7.1)	5 (1.5)	This study
▲ (as above)			10 (3.0)	
▲ (as above)			20 (6.1)	
▲ (as above)			30 (9.1)	
▼ (as above)			40 (12.2)	
▼ (as above)			54 (16.4)	
◑ Baltimore, Maryland	sand fill with brick	4.8 (5.3)	45 (13.7)	This study
● Tampa, Florida	loose sand with high water	16 (14.5)	60 (18.3)	This study
○ (as above)			20 (6.1)	
■ (as above)		7 (6.3)	60 (18.3)	
□ (as above)			10 (3.1)	
▨ Charlottesville, Virginia	silty sand rockfill	6 (5.4)	45 (13.7)	This study
⊕ (as above)			20 (6.1)	
▽ (as above)			5 (1.5)	
◆ California	silty sand fill	45 (40.5)	100 (30.5)	Gambin (7)
▲ United Kingdom	rubble fill	16.5 (15)	66 (20)	Pearce (17)
◨ Illinois	granular fill	6 (5.4)	25 (7.6)	Lukas (12)
◪ Indianapolis	granular fill	6.7 (6.0)	40 (12)	Leonards et al. (10)
◕ Seine, France	unknown	13.3 (12)	72 (22)	Leonards, et al.(10)
◒ Indiana	granular fill	15 (13.6)	60 (18.2)	Varaksin(24)
◩ Chicago, Illinois	rubble fill	3.4 (3.1)	25 (7.6)	Lukas (11)

GROUND VIBRATION DATA

Ground vibration data obtained from 12 different dynamic compaction sites were compiled during this study. Five sites were monitored by the author. The data from the other seven sites were obtained from papers and reports prepared by others (see Table I). Primarily, the sites were underlain by natural sands and granular fill materials. It is believed that the groundwater level was relatively shallow on only two of these sites: Tampa (4 feet) and Long Beach (15 feet). Data obtained by the author are expressed in terms of the true vector sum (TVS). Particle velocities reported by others are believed to be primarily in terms of single peak component or TVS. Thus, some error is introduced when comparing data of different format.

The size of weights in Table I range from 3.4 to 45 tons (3.1 to 40.5 tonnes). Drop heights varied from 5 to 100 feet (1.5 to 30.5 meters). Total theoretical energy levels per drop (WH) range between 30 to 4500 ft-tons (8 to 1235 tonne-meters). More technically correct, energy levels up to 12 MN-m were applied, however, the industry commonly uses units of tonne-meters in reporting energy levels.

Conventional crawler cranes were used to hoist and drop the weights on all 12 sites, except the Long Beach, California site, where a special tripod crane was erected. Weights were constructed of either steel or composite steel/ concrete.

The amplitude of ground vibrations attenuate with distance from the point of impact. Figure 8 presents a summary of peak particle velocity data from all sites considered. Distances as close as 7 feet and as far away as 400 feet were monitored. Based on the available data from these sites, a safe conservative upper limit (neglecting special tripod equipment) may be estimated for preliminary purposes from:

$$\text{PPV (ips)} = \left(\frac{75}{d \text{ (feet)}} \right)^{1.7} \qquad (6a)$$

$$\text{PPV (mm/sec)} = \left(\frac{153}{d \text{ (meters)}} \right)^{1.7} \qquad (6b)$$

Eq (6) does not consider the level of energy applied during dynamic compaction. Furthermore, the data is derived soley from a few sites, all underlain by granular materials. Extrapolation of these trends to sites underlain by clayey soils, variable fill materials, complex stratigraphy, shallow rock, or other dissimilarities may result in unconservative results.

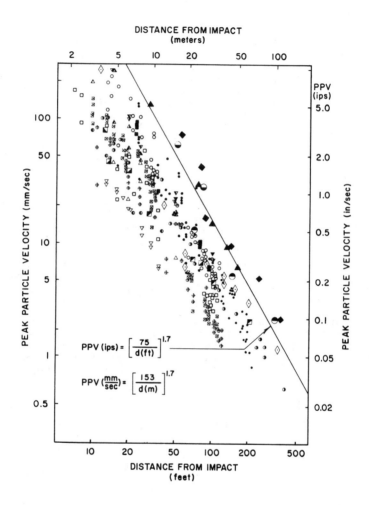

Fig. 8   Summary of peak particle velocity attenuation with
distance from impact. Data from 12 dynamic compaction
sites listed in Table I.

Within the dynamic compaction limits, it has been observed that
vibration levels increase as the treated area becomes densified (4,
17).   Generally, a maximum level of particle velocity is achieved
after one or two passes of heavy tamping or about 150 tm/m².

SCALED DISTANCE DATA

Scaled distance graphs are often used to present particle velocity data (10, 11, 15, 16, 25). Most commonly, the scaled distance axes is defined as the distance from the source to the ratio of the square-root of the applied energy. Cube-root scaling is advocated by others (1, 9). Based on the peak velocity values observed at the author's site after densification and the available supplementary data, a summary of particle velocity attenuation with inverse of square-root scaled distance is presented in Figure 9. Expressions for the upper limit of the observed trend are:

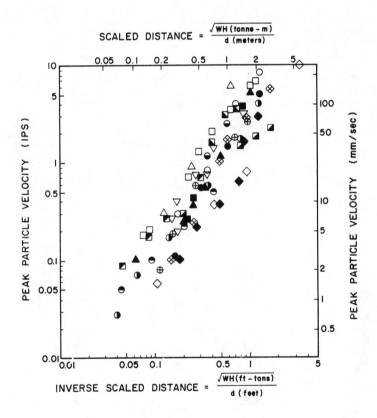

Fig. 9   Particle velocity attenuation for scaled distance according to square root of energy per blow. Data from sources given in Table I.

$$PPV \ (in/sec) \ = \ 8 \left(\frac{\sqrt{WH}}{d} \right)^{1.7} \qquad (7a)$$

where d and H are in feet and W in tons;

$$and \qquad PPV \ (mm/sec) \ = \ 92 \left(\frac{\sqrt{WH}}{d} \right)^{1.7} \qquad (7b)$$

where d and H are in meters and W in tonnes.

A close examination of Figure 9, however, indicates that the derived upper limit expression in Eq (7) appear conservative for the largest weights and highest drop heights (Tampa, Long Beach, U.K., and Alabama), possibly because of site specific differences. The effect of energy level on particle velocity was studied as suggested by Wiss (25). At several distances, the log of PPV was graphed as a function of WH, as shown in Fig. 10. Apparently, the exponent term decreases with distance away from the point of impact. At distances of 20, 50,

WH = ENERGY PER BLOW (tonne - meters)

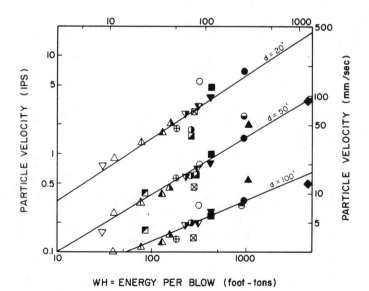

WH = ENERGY PER BLOW (foot - tons)

Fig. 10 Observed relationship between particle velocity and energy per blow at distances of 6, 15, and 30 meters.

and 100 feet, the observed effect of energy level is $(WH)^{0.6}$, $(WH)^{0.5}$, and $(WH)^{0.4}$, respectively, as determined from linear regression analyses. Such variations may be explained due to factors such as plastic deformation, material damping, geometrical damping, stratification, and other phenomena (23).

In actuality, friction in the system prevents a true free fall of the weight upon release of the clutch. The total energy per blow is somewhat less than WH. Deceleration measurements have indicated the efficiency to be on the order of 80% (8). Considering all factors involved, a more involved expression for vibration attenuation may be:

$$PPV = A_0 \ W^a \ H^b \ / \ d^c \qquad (8)$$

where A_0, a, b, and c are all parameters and not necessarily constants for a given site and the specific equipment utilized.

In an effort to discern the effects of different weight sizes on particle velocity, particle velocities were measured during dynamic compaction with a 7-ton weight and 16-ton weight in Tampa, Florida. If square root scaling applied, then the particle velocities from the 16-ton weight would be /16/7 or 1.51 times those from the 7-ton weight. For this site, the observed ratio of particle velocities averaged about 1.35, implying $(W)^{0.4}$.

The effects of drop height were investigated at a dynamic compaction site in Alexandria, Virginia. Drop heights of 5, 10, 15, 20, 30, 40, and 54 feet were monitored. At this site, the effect of drop height varied approximately as $(H)^{0.6}$ at a distance of 20 feet to $(H)^{0.4}$ at a distance of 100 feet. Based on the limited data obtained at Tampa and Alexandria, it is postulated that drop height is slightly more influential than weight size in determining the magnitude of particle velocities. Weight size may affect vibration frequency, however, as implied by equation (5).

Using an entirely new approach, the same data base from Figure 9 was re-graphed in the form of a normalized vibration level (particle velocity divided by the theoretical impact velocity of a falling weight) versus distance normalized to the weight radius (d/r_0), (see Fig. 11). Apparently, a close trend is obtained with this empirical approach, although maybe fortuitous. It would seem intuitive that the maximum possible particle velocity would occur on the weight during impact. For a free falling body, the impact velocity (v_i) is:

$$v_i = \sqrt{2 \ g \ H} \qquad (9)$$

where g = gravitational constant. In addition, for a rigid mass, the size of the weight is related to the mass radius. Referencing Figure

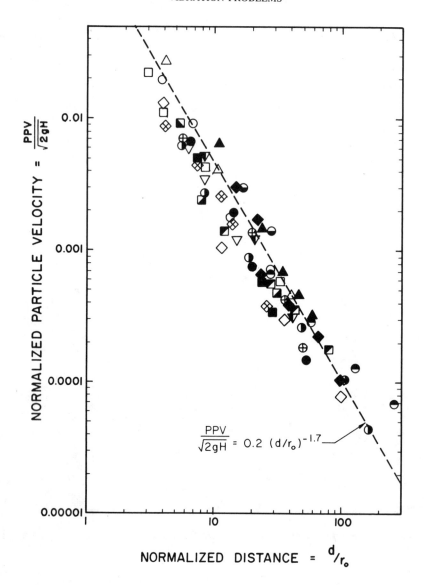

Fig. 11 Observed attenuation of normalized particle velocity with
         normalized distance. Data from sources given in Table I.

9, a linear approximation of the particle velocity attenuation using this approach would be:

$$PPV = 0.2 \sqrt{2 \, g \, H} \, (d/r_0)^{-1.7} \qquad (10)$$

where PPV and v_i are in consistent units and distance d is normalized to the radius of the weight (r_0). Since the equivalent radii of typical weights lie in the narrow range of 2 to 3.5 feet (0.6 to 1.1 m), a similiar trend is observed between normalized particle velocity and distance.

CONCLUSIONS

The findings of this study on ground vibrations caused by dynamic compaction of granular soils indicate that:

(1) Transient low-vibration frequencies resulting from dynamic compaction may require a combination velocity and displacement criterion, (or frequency-dependent velocity criterion), as recommended by two recent studies on vibration-induced damage (19, 22).

(2) It is extremely important to measure vibration frequencies of the waveform to determine whether these are within the range of the seismograph operating range. Many units are nonlinear below 6 Hz and thus, a magnification factor may be required.

(3) Vibration levels should be reported in terms of maximum single component amplitude or true vector sum (TVS), not pseudo vector sum (PVS).

(4) Particle velocity attenuation from square-root scaled-distance graphs appear applicable for data from 12 different sites underlain by granular soils and reviewed in this study. Significant trending was also observed for particle velocities normalized to impact velocity and attenuated with distance normalized to the weight radius.

(5) Additional research and data are needed on the use of spectrum analyses for vibration monitoring during dynamic compaction. In addition, a damage criterion based on dynamic compaction data is warranted.

ACKNOWLEDGEMENTS

The writer appreciates the help of Anne Bethoun, Donna Reese, and Randy Galpin in preparation of this manuscript. Thanks to Alan Foster of Vibratech Engineers and Michel Gambin of Soletanche for reviewing the text.

REFERENCES

1. Ambrasseys, N. and Hendron, A.J., "Dynamic Behavior or Rock Masses", Rock Mechanics in Engineering Practice, Editor - Stagg, John Wiley & Sons, London, 1968, pp 203-227.

2. Chae, Y.S., "Design of Excavation Blasts to Prevent Damage," ASCE Civil Engineering, April 1978, pp 77-79.

3. Dobson, T and Slocombe, B., "Deep Densification of Granular Fills," paper presented at the 2nd Geotechnical Conference, ASCE National Convention, April 1982, Las Vegas, Nevada, 11 pages.

4. Dumas, J.C., "Dynamic Consolidation: Case Histories, Canadian Realizations," 1982 Reports by Geopac, Inc. 680 Birch, St.-Lambert, Quebec, J4P 2N3.

5. Dumas, J.C. and Beaton, N.F., Limitations and Risks of Dynamic Compaction" paper presented at the ASCE National Convention, Atlanta, Georgia, May 16, 1984, 10 pages, Geopac, Inc.

6. Foster, G.A., "Blasting Vibration Control - A Time of Change," Stone News National Crushed Stone Association, August 1981, pp 19-24.

7. Gambin, M.P., "Menard Dynamic Compaction," Proceedings of ASCE National Capital Section Seminar on Ground Reinforcement, George Washington University, Washington, D.C., January 1979.

8. Hansbo, S., "Dynamic Consolidation of Rockfill," Proceedings 9th International Conference on Soil Mechanics and Foundation Engineering, Vol. 2, Tokyo, 1977, pp 241-246.

9. Hendron, A.J., "Engineering of Rock Blasting on Civil Projects," Proceedings ASCE National Capital Section Seminar Recent Trends in Geotechnical Engineering Practice, George Washington University, February 5, 1982, 57 pages.

10. Leonards, G.A. et al, "Dynamic Compaction of Granular Soils," Journal of the Geotechnical Engineering Division, Vol. 106, No. GT 1, January 1980, pp 35-44.

11. Lukas, R.G., "Densification of Loose Deposits by Pounding," Journal of Geotechnical Engineering, Vol. 106, No. GT 4, April 1980, pp 435-446.

12. Lukas. R. G., "Dynamic Compaction Manual," STS Consultants Report to Federal Highway Administration, Washington, D.C., Contract No. DTFH-61 -83-C-00095, March 1984.

13. Martin, D., "Ground Vibrations from Impact Pile Driving," Transport and Road Research Laboratory Report 544, 1980, Crowthorne, England, 25 pages.

14. Mayne, P.W. and Jones, J.S. "Impact Stresses During Dynamic Compaction" Journal of Geotechnical Engineering, Vol. 109, NO. 10, Oct. 1983, pp 1342 -1346.

15. Mayne, P.W., Jones, J.S. and Dumas, J.C., "Ground Response to Dynamic Compaction," Journal of Geotechnical Engineering, Vol. 110, No. 6, June 1984, pp 757-774.

16. Mitchell, J.K., "Soil Improvement: State-of-the-art Report," Proceedings, 10th International Conference on Soil Mechanics and Foundation Engineering, Vol. 4, Session 12, Stockholm 1981, pp 509-521.

17. Pearce, R.W., "Deep Soil Compaction by Heavy Surface Tamping," State-of-the-art report in partial fullfillment for Master of Science degree, Imperial College, University of London, July 1977, 150 pages.

18. Richart, F.E., Hall, J.R., and Woods, R.D., Vibrations of Soils and Foundations, Prentice-Hall, New Jersey, 1970, 414 pages.

19. Siskind, D., Stagg, M. et al, "Structure Response and Damage Produced by Ground Vibration from Surface Mine Blasting," Bureau of Mines Report RI 8507, 1980, Twin Cities, Minnesota, NTIS #PB81-157000, 74 pages.

20. Skipp B., and Buckley, J., "Ground Vibration from Impact," Proceedings 9th International Conference on Soil Mechanics and Foundation Engi neering, Vol. 2, Tokyo, 1977, pp 397-400.

21. Stagg. M. and Engler, A., "Measurement of Blast Induced Ground Vibrations and Seismograph Calibration," Bureau of Mines Report RI 8506, 1980, Twin Cities, Minnesota, NTIS #PB81-157828, 62 pages.

22. Studer, J. and Suesstrunk, A., "Swiss Standard for Vibrational Damage to Buildings," Proceedings 10th International Conference on Soil Mechanics and Foundation Engineering, Vol. 3, Stockholm, 1981, pp 307-312.

23. Taniguchi E., and Okada, S., "Reduction of Ground Vibrations," Soils and Foundations, Vol. 21, No. 2, June 1981, pp 99-113.

24. Varaksin, S.V., "Report on Soil Improvement for Indianapolis Public School No. 47," Menard Project No. 79-008, January 11, 1980, Pittsburgh, Pennsylvania

25. Wiss, J.F., "Construction Vibrations: State-of-the-art," Journal of the Geotechnical Engineering Division, Vol. 107, No. GT2, February 1981, pp 167-182.

INVESTIGATION OF POTENTIAL DETRIMENTAL
VIBRATIONAL EFFECTS RESULTING FROM
BLASTING IN OILSAND

B.R. List, E.R.F. Lord, A.E. Fair*

Syncrude Canada Limited operates an open pit oilsand mine in northeastern Alberta, Canada. Oilsand production of approximately 160,000 m^3 per day is achieved utilizing draglines to excavate the oilsands in a single bench mining operation with up to sixty (60) metre high cuts. Bitumen is extracted from the oilsand and upgraded to produce an average of approximately 110,000 barrels of synthetic crude oil per day.

The depositional environment of the oilsand is such that certain areas require remedial stabilization to ensure the safety of the draglines. These areas result in a greater risk to the operation of a dragline close to the highwall crest unless the zone containing steeply dipping clay beds can be stabilized. One of two techniques currently utilized involves preblasting the area in advance of mining. An increase in stability is achieved by both disrupting the overall continuity and reducing the pore water pressures of the clay beds.

One major concern that had to be examined was the potentially detrimental vibrational effects in adjacent unblasted areas. Studies were undertaken to determine both the extent of any adverse effects beyond the blast perimeter and any related reductions in soil strength. The study utilized several field instrumentation techniques to measure various parameters resulting from a blast. It also involved a series of laboratory experiments to determine the effects of cyclical loading on the soil strength.

Field monitoring of the blasts has indicated that potential detrimental vibrational effects in adjacent unblasted areas are limited. Based on these studies, guidelines have been developed and are currently being utilized for all remedial stabilization blasting. The studies were initiated in 1980 and have continued by utilizing subsequent blast programs to expand the data base.

INTRODUCTION

Syncrude Canada Limited operates an open pit oilsand mine in northeastern Alberta, Canada. Oilsand production of approximately 160,000 m^3 per day is achieved utilizing draglines to excavate the oilsands in a single bench mining operation. Bitumen is extracted from the oilsand and upgraded to produce an average of approximately 110,000 barrels of synthetic crude oil per day.

*Geotechnical Engineering Section, Syncrude Canada Limited, Fort McMurray, Alberta, Canada

The mining method used at the Syncrude mine site as shown in Fig. 1 utilizes a single bench dragline scheme with open pit highwall depths of as much as sixty (60) metres. During the oilsand excavation large "block slides" involving the upper portion of the highwall have occasionally occurred along the newly excavated slopes. In all cases, the controlling factor was the presence of clay beds dipping out of the highwall slopes. The possibility of block slides continues to pose the largest single threat to overall highwall stability and dragline safety.

Several techniques are currently in place to both identify and monitor potential block slides. Normally areas which exhibit a higher potential for highwall instability are dealt with by increasing dragline tub setbacks or by using constraints on the direction of mining. In areas where clay beds, on which large blocks can slide, dip steeply straight out of the highwall, directional mining constraints will not solve the problem. When this situation occurs remedial stabilization techniques are required to ensure highwall stability. To date, two remedial stabilization techniques have been developed and successfully implemented to deal with such situations. The two techniques are sub-excavation/backfilling, and blasting.

The sub-excavation/backfilling technique requires that the area of potential instability be excavated, usually by dragline, then backfilled, generally with the same material hence disrupting the clay beds. This technique was first used in 1980 and has since been successfully used in several areas of the east mine. The looser material state that is created does not yield stability problems but does result in some settlement due to loading. This procedure represented a significant step forward as previous buttresses, which involved considerable non-productive dragline time had been utilized. However, the sub-excavation/backfilling technique typically requires two (2) to four (4) days of dragline time to complete the necessary work for one problem area. As a result, additional work in the area of remedial stabilization methods led to the development of the preblasting technique.

Blasting for the purpose of stabilizing an oilsand highwall is without precedent. Blasting is conducted for the purposes of disrupting clay beds on which large blocks could potentially slide and also to lower the pore water pressures within the oilsand. Initial development work was undertaken with the assistance of Dr. A. Bauer of Queen's University. As this was the first application of explosives to remedial stabilization, extensive monitoring was utilized to assess the success of blasting. A laboratory study into the effects of the cyclic loading and subsequent potential reduction in shear strength was also undertaken at the University of British Columbia.

Of major concern in these blast programs is the blast generated vibrations; both vibrations through the ground and through the atmosphere. The prime concerns are centered around the potential reduction of the soil strength in adjacent areas due to the vibrations. The effects that the blast induced vibrations have on the near-by processing plant are also of prime concern. Other blasting concerns such as flyrock and interference with the normal mining operation are inherent with any blasting operation and are not addressed in this paper.

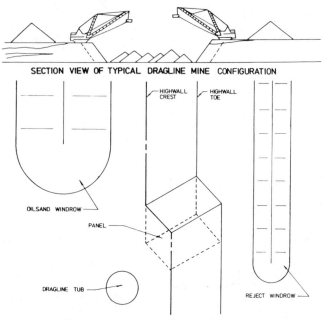

SECTION VIEW OF TYPICAL DRAGLINE MINE CONFIGURATION

SCHEMATIC PLAN OF DRAGLINE CUT – GEOMETRY

Figure   1   Syncrude Dragline Mining Method

GEOLOGY AND HIGHWALL STABILITY

In order to understand the origin of adversely dipping clay beds, a basic knowledge of Syncrude's mine geology is required. The McMurray Formation, which contains the oilsand deposit, is the sediment fill of a basin developed on eroded Devonian carbonates (Waterways Formation) during the Pre-Cretaceous transgressive sequence. The various sedimentary facies reflect successive deposition in fluvial, estuarine and marine environments. This type of depositional setting can be broadly termed as 'tidal deposition'. Tidal flats develop along gently dipping sea coasts with established tidal rhythms, where enough sediment is available for deposition and strong wave energy is not present. The surface of the tidal flat sediment slopes gently towards the sea from the high-water level toward the low-water level. As a result, tidal flat sediments have a gentle initial dip, corresponding to the seaward sloping surface, of a few degrees.

Drainage of the intertidal flats was accomplished by a network of interconnected meandering tidal channels which continuously reworked the tidal flat area. Deposition of sand, silt and/or clay occurs along the convex side of the channel resulting in 'point bars' which consists of a series of inclined layers. The inclined beds produced

by the lateral migration of channels dip in a direction normal to the
channel trend. The lower portion of the inclined bed sequences are
tangential to the basal scoured surface. These inclined beds can dip
up to twenty degrees or more normal to the channel. The lateral
continuity of the dipping beds will be dependent on the degree of
meandering. Fig. 2 illustrates the development of typical series of
inclined layers within a tidal flat environment.

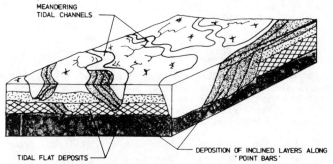

DEVELOPMENT ON INCLINED LAYERS IN A TIDAL FLAT ENVIRONMENT

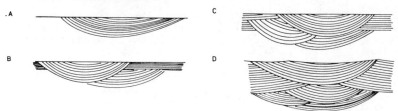

Figure   2   Development of Tidal Channel - Fill Cross-Bedding

The presence of continuous inclined clay layers is the main factor
leading to the occurence of block slides. The regional dip toward the
west results in a strong westerly dip direction for the inclined clay
layers formed by the meandering tidal channels. As the Syncrude mine
is orientated in an approximate north/south direction, the east
highwall as opposed to the west highwall has the highest risk of block
slides.

A second factor involved in the occurrence of block slides is the
shear strength of the clay layers. Shear strength testing of both the
tidal channel mud and bedded shoreface facies, indicates a peak
effective shear strength of 13^0 to 29^0, and a residual strength of
8^0 to 25^0 (1). The clay layers are normally interbedded with sand
and silt layers and are generally very thin, usually less than one
centimetre in thickness. Horizontal displacement along these clay
layers, causing a subsequent reduction in shear strength, results from
both stress relief due to mining of the adjacent panel and the induced
stresses resulting from both the dragline and windrow loading on the
highwall bench.

The third factor related to the occurrence of block slides is the elevation of the piezometric surface acting on potential failure surfaces. The high remnant pore water pressures in the area of the Syncrude mine are a result of the loading conditions during the last ice age. Due to the low hydraulic conductivities of both the clay layers and overall oilsand, the excess pore pressures within the thin clay layers, which form the potential failure surfaces, are not able to dissipate during the relatively short time between the mining of successive highwall cut panels.

BLAST PROGRAM OBJECTIVES AND DESIGN

With the current mining method employed at Syncrude it is essential that highwall stability be maintained. There are four factors which greatly effect the stability of the highwall and this dragline safety. These factors are geological conditions, pore water pressures, loading conditions and stress relief.

Through a combination of these factors, a block slide may occur. An area, which based on geological information, may place a dragline in a high risk situation must undergo remedial stabilization. This has come in the form of sub-excavation and backfilling or more recently through blasting.

With the expected increase in plant production and requirements from the mine, an ongoing remedial stabilization program is expected in the future. The objectives of remedial stabilization blasting are as follows:

1. Disruption of steeply dipping clay beds normal to the channel axis

2. Reduction of high pore water pressures

3. Minimization of damage to the highwall slope adjacent to the blast area

4. Minimization of detrimental blast induced vibrational effects to adjacent areas not undergoing remedial stabilization.

The failure plane on which a block slides in most cases is a relatively weak, thin steeply dipping clay layer. Disruption in continuity of these clay layers results in greater friction along the potential failure surface and hence increases the factor of safety. The magnitude of disruption need not be great for improvement in the overall highwall stability.

Theoretical analysis and past performance has shown that the greatest improvement in highwall stability comes from the reduction in high pore water pressures within the blast areas. This is done by providing the material with short and direct drainage paths. Micro fractures produced by the blast action are sufficient to reduce the pore water pressures within the oilsand. The micro fractures are also thought to yield smaller changes in pore water pressures due to external loading.

Although the prime objective of a blast is to disturb material, within the problem area, this is not desired along the highwall face. Excessive kick-out during a blast may reduce the efficiency of the mining operation. The blast area must therefore be sufficiently offset from the highwall crest, or the blast designed accordingly to minimize damage to the highwall face.

A major objective of highwall stabilization blast programs is to minimize the disturbance to adjacent unblasted areas and that no additional potentially unstable areas are created by reducing the soil strength. An extensive monitoring program including direct and indirect monitoring of the vibrational effects on the oilsand and on highwall stability was incorporated into the blast programs. A laboratory study into the effects of cyclic loading and subsequent potential reduction in material shear strength was also undertaken at the University of British Columbia.

The blast program objectives are achieved by vertically displacing the oilsand utilizing spherical or squat charges in a crater blasting mode. Research into the effects of crater blasting by Dr. Bauer established a plot of the true crater volume per unit charge weight versus the scaled depth of burial of the explosive charge. The "scaled depth of burial" of an explosive charge is used for blasthole design and for comparison purposes between different designs. It represents a relationship between the depth to the center of the explosive charge and the cube root of the explosive weight as follows:

$$SDB = \frac{d}{W^{1/3}} \qquad (1)$$

where: SDB = Scaled depth of burial
          d = Depth to center of charge
          W = Explosive weight

It was found that the optimum crater is produced at a scaled depth of burial of 3.0 in overburden material and the radius of the crater is approximately equal to the depth of placement. Testing done at Syncrude found the optimum crater to be produced at a scaled depth of burial of approximately 3.2 for oilsand. At the free surface, or the mine bench, the crack limit radius was found to be double that of the scaled depth of burial of the explosive charge. The crater blasting design parameters are shown in Figs. 3 and 4.

At Syncrude, the most cost efficient method of remedial stabilization blasting of this type is with the use of AN/FO (Ammonium Nitrate and fuel oil) in large diameter blastholes. The current remedial stabilization blast programs utilize blastholes of 30 in. (0.76m) in diameter, and approximately 50 ft. (15m) in depth. The explosive weight per blasthole is approximately 2600 pounds (1180 kg). By limiting the amount of explosives per delay period and utilizing a variety of delays between individual blastholes the peak particle velocity due to blasting is controlled. The typical blasthole design used is shown in Fig. 5.

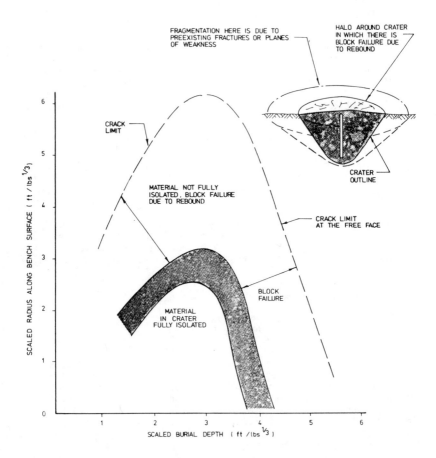

Figure   3   Crater Blasting Design[5]

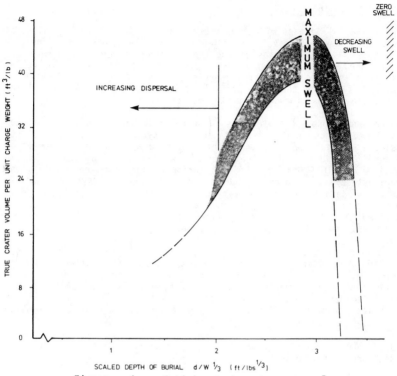

Figure   4   Optimium Crater Blast Design[5]

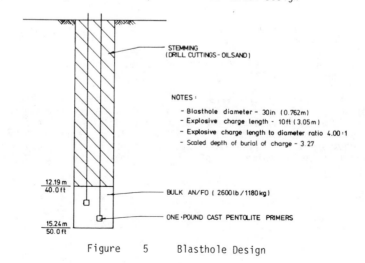

NOTES :
  - Blasthole diameter - 30in (0.762m)
  - Explosive charge length - 10ft (3.05m)
  - Explosive charge length to diameter ratio 4.00 : 1
  - Scaled depth of burial of charge - 3.27

STEMMING
(DRILL CUTTINGS - OILSAND)

BULK AN/FO (2600 lb /1180 kg)

ONE·POUND CAST PENTOLITE PRIMERS

12.19 m
40.0 ft

15.24 m
50.0 ft

Figure   5   Blasthole Design

DETRIMENTAL SIDE-EFFECTS

The principal factors effecting vibration levels at a given geological
site are the explosive weight fired per delay, distance from the
blast, and the properties of the transmitting medium. In engineering
there are generally accepted peak particle velocity threshold values
at which certain types of damage start to occur as illustrated in
Table I (2). As the peak particle velocity is increased beyond the
threshold values, the likelihood of damage also increases.

Bauer and Crosby (3) established a relationship in oilsand between
scaled distance and peak particle velocity for single hole blasts
using a combination of measurements obtained at Syncrude and previous
Bauer/Crosby data. This relationship was modified by Bauer (4) to
include multiple hole blasts. Vibration measurements recorded in the
earliest blast program conducted at Syncrude in 1980 fitted the
previous data well and it was concluded that the anticipated peak
particle velocity at a given distance for multiple hole blasts is
defined by the upper limit line and could be used for design
purposes. This design curve is shown in Fig. 6.

By design, this line is conservative for use in design as there is a
considerable scatter of data which can amount to differing values
approaching an order of magnitude. To use the line for a more
efficient blast design it was necessary to establish the critical peak
particle velocity which the adjacent unblasted area could sustain
without detrimental effects. An alternative approach is to establish
a critical scaled distance beyond which there is no effect.

The studies currently underway have been divided into three aspects
which are as follows:

1. To verify the upper limit line and to establish a lower limit
    line based on the blasthole design currently being used.

2. To identify the extent over which lateral movement within the
    oilsand occurs.

3. Having identified the critical zone, what if any, is the
    reduction in soil strength.

The design curves were established using velocity recorders which
monitor the blast generated vibrations directly. The extent over
which blast generated movement occurs was identified by the use of
slope inclinometers. Changes in soil strength was examined through a
laboratory study into the effects of cyclic loading by the Engineering
Department at the University of British Columbia.

TYPE OF STRUCTURE	TYPE OF DAMAGE	PEAK PARTICLE VELOCITY THRESHOLD AT WHICH DAMAGE STARTS (IN / SEC)
RIGIDLY MOUNTED MERCURY SWITCHES	TRIP OUT	0.5
HOUSES	PLASTER CRACKING	2 ← SET INITIAL LIMIT OF 5 IN/SEC MAXIMUM AT THE CRUSHER
CONCRETE BLOCK AS IN A NEW HOUSE	CRACKS IN BLOCKS	8
CASED DRILL HOLES RETAINING WALLS LOOSE GROUND	HORIZONTAL OFFSET	15
MECHANICAL EQUIPMENT PUMPS, COMPRESSORS	SHAFTS MISALIGNED	40 BEYOND 10 IN/SEC MAJOR DAMAGE STARTS SUCH AS POSSIBLE CRACKING OF CEMENT BLOCK
PREFABRICATED METAL BUILDING ON CONCRETE PADS	CRACKED PADS BUILDING TWISTED AND DISTORTED	60

Table   1   Type of Damage Related to the Peak Particle Velocity in the Ground Waves from Blasts[2]

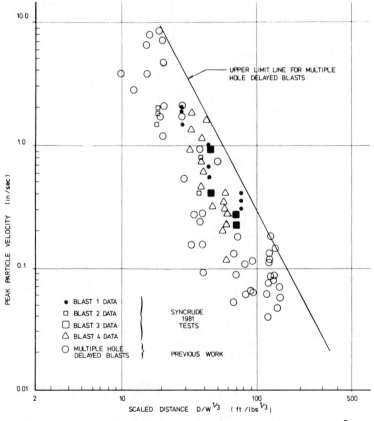

Figure   6   Peak Particle Velocity Vs. Scaled Distance[5]

Ground Particle Velocity Monitoring

A velocity recorder is used to monitor the blast induced vibrations (longitudinal, vertical and transverse modes) within the soil mass at set distance from a blast. The instrument, if placed on the ground surface, measures the peak surface particle velocity in inches per second. This value is then converted to a peak particle velocity by dividing the obtained reading by 2.0.

The use of large diameter blastholes with large explosive weights required that the resultant blast generated vibration be examined and that their compliance with the design curve be verified. Monitoring of the highwall stabilization blasting conducted in 1984 showed that the upper limit line for multiple hole blasts previously established is valid for the new blast design. The monitored vibration levels in the various modes (longitudinal, vertical and transverse) are shown in Fig. 7, and tabulated in Table 2. Monitoring of the blast vibrations revealed that the transverse waves were consistently lower than the vertical and longitudinal waves. Based on the data obtained during the 1984 blast program a lower limit line was also established as shown in Fig. 7.

Extent of Blast Generated Movement in Oilsand

Slope inclinometers have been successfully used to monitor ground response due to blasting. Vertical slope inclinometers were installed to determine subsurface lateral movements in the adjacent unblasted areas. For blast induced movement, a minimum amount of movement used is 0.06 in. (1.52mm) in either of the two instrument axes. The slope inclinometer consists of a grooved casing grouted into a borehole to a depth well below the depth of the blast or depth of anticipated movement. The grooves control the orientation of the sensor which is lowered to the bottom of the hole. It is pulled back up at 2 ft. (0.6m) intervals and readings in the x and y direction are taken before and after each blast, to determine whether there has been any measurable ground response.

Slope inclinometers within a radius of a scaled distance of approximately twenty-five that were previously installed to monitor ground movement during mining were used to monitor blast induced ground response. The data tabulated in Table 3 represents slope inclinometer readings obtained by instruments which had shown no prior movement, either through initial blast response or highwall instability unrelated to blasting.

Based on the data shown in Table 3 and plotted in Fig. 8, the critical scaled distance is at approximately 7.4. This point represents the distance away from the nearest blasthole to which lateral movement can be detected through the use of slope inclinometers.

With the exception of three instrument readings, all readings conformed to the critical scaled distance of 7.40. Slope Inclinometer SI 840210 shown no movement at a scaled distance of 6.7. Slope inclinometer SI 840037 showed minimal movement of 1.97 millimetres in

TABLE 2: VIBRATION DATA - HIGHWALL STABILIZATION BLASTING - 1984

BLAST NUMBER	SCALED DISTANCE TO NEAREST BLASTHOLE	PEAK PARTICLE VELOCITY (IN/SEC)		
		LONGITUDINAL WAVE	VERTICAL WAVE	TRANSVERSE WAVE
84-1/1	25.10	2.225	2.00	0.85
	25.10	2.50	2.225	1.25
84-1/2	25.70	2.05	2.60	1.10
	25.70	2.125	3.00	1.00
84-2/1	18.77	1.75	3.75	0.75
	32.84	0.875	1.175	0.45
84-2/2	18.52	2.25	2.15	1.15
	18.52	2.25	2.25	1.40
84-2/3	19.07	2.15	2.40	1.50
	16.82	2.40	3.50	1.50
84-2/4	18.77	2.90	5.75	2.15
	15.96	4.15	8.25	1.75
84-2/5	14.07	5.75	4.15	2.50
	14.99	4.50	4.25	2.50
84-2/6	14.45	8.25	7.15	3.50
	15.66	5.25	4.65	3.15
84-2/7	16.60	2.90	3.90	3.75
	16.88	2.65	3.75	3.15
84-3/1	11.74	11.50	9.30	3.50
	13.23	8.30	9.00	2.80
84-3/2	10.84	5.30	10.50	4.30
	11.28	5.80	10.50	3.80

response to blasting. Movement of 6.54 millimetres at a depth of sixty-four feet, well below the extent of blasting, identified by Slope Inclinometer SI 840060 can be attributed to an inherently weak layer below the blast area and may not be attributed to the blasting action alone.

The response to blasting as identified by the slope inclinometers was plotted on a graph comparing the scaled distance to the peak particle velocity. The data was plotted on the conservative lower limit line thus providing a minimum peak particle velocity that would yield lateral movement within material in areas adjacent to a blast area. The critical scaled distance of 7.4 corresponds to a minimum peak particle velocity of approximately 8.9 in./sec. before lateral movement will occur.

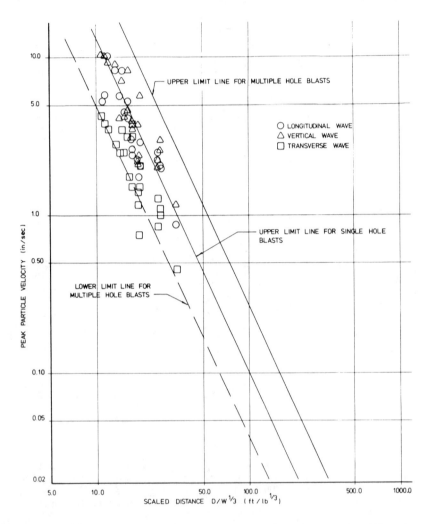

Figure   7   Peak Particle Velocity Vs. Scaled Distance for Oilsand
                 (1984 Blast Program)

TABLE 3: SLOPE INCLINOMETER DATA - HIGHWALL STABILITY
BLASTING - 1984

SLOPE INCLINOMETER NUMBER	BLAST NUMBER	SCALED DISTANCE TO NEAREST BLASTHOLE	MAXIMUM RESPONSE MOVEMENT (mm)
SI 830216	84-1/1	2.75	Destroyed
SI 830218	84-1/1	7.27	5.22mm
SI 840032	84-1/1	10.10	No movement
SI 840033	84-1/1	5.72	8.17mm
SI 840034	84-1/1	8.88	No movement
	84-1/2	9.31	No movement
	84-1/3	15.70	No movement
SI 840035	84-2/1	12.26	No movement
	84-2/2	4.64	7.24mm
SI 840036	84-2/1	3.13	26.36mm
SI 840037	84-1/1	15.58	No movement
	84-1/2	9.11	1.97mm
SI 840038	84-1/2	18.09	No movement
	84-1/3	11.25	No movement
	84-2/1	9.79	No movement
SI 840039	84-1/2	15.18	No movement
	84-1/3	7.07	5.41mm
SI 840040	84-1/2	13.06	No movement
	84-1/3	6.79	9.03mm
SI 840041	84-1/2	12.54	No movement
	84-1/3	5.45	6.24mm
SI 840043	84-2/5	20.10	No movement
	84-2/6	11.05	No movement
	84-2/7	5.65	7.08mm
SI 840060	84-2/3	21.48	No movement
	84-2/4	16.75	No movement
	84-2/5	12.62	6.54mm
SI 840100	84-3/1	13.74	No movement
	84-3/2	23.85	No movement
SI 840112	84-3/1	14.68	No movement
	84-3/2	16.33	No movement
SI 840116	84-3/1	7.36	2.74mm
SI 840117	84-3/1	7.43	No movement
	84-3/2	10.63	No movement
SI 840118	84-3/1	4.96	9.33mm
SI 840125	84-3/1	5.40	17.54mm
SI 840209	84-3/1	7.77	No movement
SI 840210	84-3/1	12.69	No movement
	84-3/2	6.72	No movement

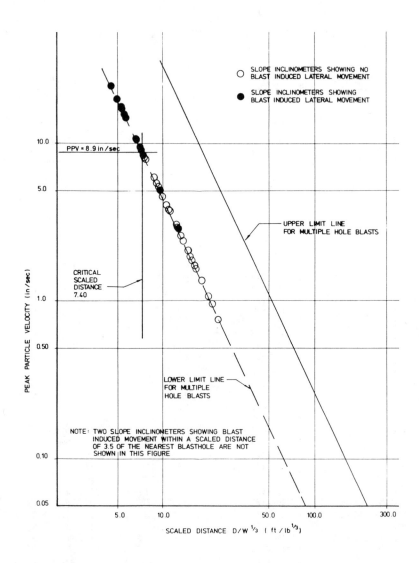

Laboratory Testing Program

A laboratory testing program involving a series of both static and cyclic triaxial compression tests was undertaken to examine the behaviour of adjacent unblasted materials in response to blasting. Both core and block samples were provided to the University of British Columnbia, where the testing was completed (6).

Particle velocity data was used to estimate the shear strains caused by blasting in the adjacent areas. The following relationship illustrates the parameters utilized:

$$\gamma = \frac{v}{c} \tag{2}$$

where: γ = shear strain
         v = particle velocity
         c = wave propagation velocity

Data obtained by Bauer (5) and subsequently by Syncrude, during additional blast programs provides an upper limit line for multiple hole blasts allowing the particle velocity to be calculated for various charge weights and scaled distances. Measured wave propagation velocities in tarsand range from 4000 to 8000 ft/sec (3). Shear strain calculations were based on a value of 4000 ft/sec. as a low estimate of wave speed will produce a conservative estimate of the shear strain. Based on numerous peak particle velocity measurements, at various scaled distances, and subsequent visual/field observations it appears that up to scaled distances of about 6 tension failures and expansion can be expected. At a scaled distance of 6, the peak particle velocity is 35 in/sec. based on the upper limit line. Using the earlier relationship with c = 4000 ft/sec.:

$$\gamma = \frac{v}{c} = \frac{35}{4000 \times 12} = 0.0007 = 0.07\% \cdot \tag{3}$$

Consequently, based on the above data and observations the material in the adjacent unblasted zone can be expected to be subjected to cyclic shear strains of up to 0.07%.

The testing program was specifically intended to determine the effect of the blast induced stresses on both the strength of the material and the pore fluid pressures. A typical blast has a duration of approximately 2 seconds which results in about 20 to 30 stress cycles based on a dominant frequency of about 10 to 12 cps (3).

Prior to commencing cyclic triaxial testing a series of static tests were carried out to determine the pre-cycled strength. The samples were prepared such that failure would be initiated along bedding planes by coring the material at an angle of 45 - $\emptyset/2$ to the axis of the sample. Subsequent cyclic loading results were than compared to the static results in terms of post-cycled reduction in strength with the level of cyclic stress or strain. The cyclic tests were performed in stages covering a cyclic strain ranging from 0.25% to 4%, using

twenty to fifty cycles of stress. A loading frequency of 1 cps rather
than 10 cps was used as the UBC cyclic test apparatus was best suited
to 1 cps. This was considered acceptable as pore pressure response
for both sands and clays is governed by cyclic strain rather than
stress.

Fig. 9 illustrates the results in terms of the variation in strength
with the level of cyclic shear strain. The results indicate that the
pre-cycled or static effective friction angle, $\emptyset$, equals approximately
34° (cyclic strain = 0), and that the post-cycled $\emptyset$ is not
significantly reduced by cyclic strain as high as 4%.

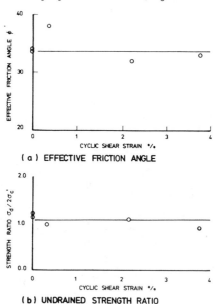

(a) EFFECTIVE FRICTION ANGLE

(b) UNDRAINED STRENGTH RATIO

Figure   9   Effective Friction Angle and Undrained Strength Ratio
                 Vs. Cyclic Strain[6]

The pore pressures induced by the cyclic strains are illustrated in
Fig. 10. An approximate upper bound line suggests that pore pressures
begin to develop at shear strain levels of about 0.01%, and reach a
peak pore pressure ratio of 0.23 at 1 to 2% shear strain. At 0.07%
strain the pore pressure ratio is about 0.13. Beyond a distance of
about 130 feet no increase in pore fluid pressure is expected to occur.

The laboratory testing program suggests that the major adverse effect
of the remedial blasting program is a predicted slight increase in the
pore water pressure ratio, r_u, equivalent to $r_u \doteq 0.1$. Additional
blast monitoring programs have and will be directed, in part, to
obtain pore water pressure data to verify the extent of any change in
r_u. No appreciable reduction in the shear strength of the adjacent
unblasted material is expected based on the laboratory results.

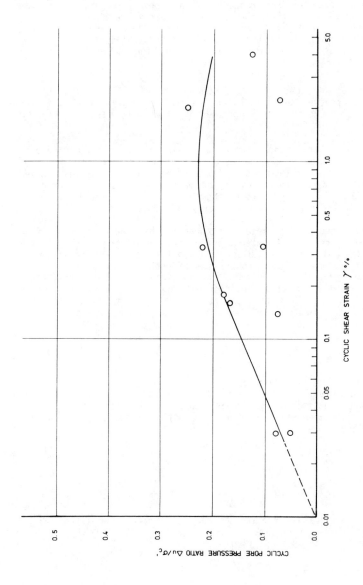

Figure   10   Pore Pressures Induced by Cyclic Shear Strain6

CONCLUSIONS

The increase in oilsand requirements by the processing plant requires that the majority of the remedial stabilization in the future be conducted by blasting to minimize non-productive dragline time. Post blast monitoring and successful mining through blast areas has shown that blasting is a viable option for remedial stabilization. A prime concern of blasting is the resultant vibrations, through both the ground and the atmosphere.

Laboratory testing conducted to determine the effect of blast induced oscillating stresses and strains on the soil strength and on the pore fluid pressures has shown the detrimental effects to adjacent areas to be minimal.

The extent of lateral movement from a blasthole, as determined through the use of slope inclinometers, is to a scaled distance of approximately 102 ft. (31m) using the current blasthole design. Based on the data obtained during the remedial stabilization blast programs in 1984, lateral movement within the oilsand is expected to occur in regions where the peak particle velocity exceeds 8.9 in/sec.

The data obtained during these blast programs has therefore identified a region of concern which extends approximately 102 ft. (31m) from the nearest blasthole. Crater blasting theories developed as shown in Fig. 3 dictate that block failure due to rebound occurs to a scaled distance twice that of the optimum scaled depth of burial. This limit therefore is to a scaled distance of 6.4; an actual distance based on current design practice of 81.0 ft. (26.8m). Similarily, material within a crater is fully isolated to a scaled distance of 3.2; an actual distance based on current design practice of 40.5 ft. (13.4m). The area in which there could be detrimental side effect therefore ranges between approximately 40.5 ft. (13.4m) and 81.0 ft. (31.0m) from a blasthole.

The determination of blast area limits therefore incorporates the above finding and assures that the potentially detrimental side effects are restricted to areas that are deemed very stable. Additional instrumentation will also be put in these areas to assure safe dragline mining.

ACKNOWLEDGEMENTS

Many sections have contributed to the development and monitoring of the current blast programs. Messrs. B. Muir, C. O'Leary, and B. Rutherford are thanked in particular.

1. Brooker, E.W. and Khan F. 'Design and Performance of Oilsand Surface Mine Slope' Oil Sand Geoscience Conference, Edmonton, Alberta 1980.

2. Bauer A., Calder, P.N. Crosby W.A., and Workman L. 'Drilling and Blasting in Open Pits and Quarries' Course Notes 1982, Parts C and D.

3. Bauer A. and Crosby W.A. 'The Effect of Blast Vibrations on Tar Sand Slope Stability II' Report for Syncrude Canada Ltd. 1976.

4. Bauer A. 'Tarsand Test Blast Designs for Syncrude Canada Limited' Report for Syncrude Canada Ltd. 1980.

5. Bauer A. 'Blast Designs to Produce Improved Highwall Stability for Syncrude Canada Ltd.' Report for Syncrude Canada Ltd. 1981.

6. Anderson D.L. Byrne P.M., and Finn W.O. 'Effect of Controlled Blasting on Highwall Stability, Syncrude Project' Report for Syncrude Canada Ltd. 1983.

DYNAMIC ANALYSIS AND PERFORMANCE OF COMPRESSOR FOUNDATIONS

Krishen Kumar[1], Shamsher Prakash,[2]F.ASCE, M.K.Dalal[3], and R.K.M. Bhandari[4]

ABSTRACT: Vibration and settlement characteristics of two identical massive concrete block foundations for 600 rpm reciprocating compressors have been studied. Large foundation settlements and frequent misalignment of the shafts were noticed so that trial runs could not be taken for two years. Preloading to 44% overload, free vibration and steady state forced vibration tests in various directions were carried out on one of the blocks to determine static and dynamic soil properties. Several test runs at different machine loads were conducted on another foundation, monitoring the vibration amplitudes, settlements, and misalignments. The performance of the foundations was satisfactory, thus warranting trial and regular runs on all three foundations.

DESCRIPTION OF COMPRESSORS AND FOUNDATIONS

Three air compressor foundations in a Refinery had undergone excessive settlements (EIL Report, 1983). The clearance between foundations is 3.38 m (11.1 ft). These were cast on 26.12.1980, 17.01.1981 and 08.01. 1981 in the sequence A, B and C respectively. The air compressors are two stage (L.P. and H.P.) reciprocating machines. Each unit comprises of two banks of two crank-throw balance opposed compressors and has a 500 KW 595 RPM induction motor with double shaft extension to drive the two banks from either end through torsionally rigid-flexible metal membrane couplings. Each compressor bank supports one L.P. (1st stage) and one H.P. (2nd stage) cylinder on either side of it as shown in Fig.1.

---

1 Professor in Civil Engineering, University of Roorkee, Roorkee,India.

2 Director, Central Building Research Institute, Roorkee, India.

3 Project Co-ordinator, Engineers India Ltd., PTI Building, Sansad Marg, New Delhi, India.

4 Supervising Engineer, Engineers India Ltd., PTI Building, Sansad Marg, New Delhi, India.

Placed underneath the two L.P. cylinders is the common 1st stage suction volume bottle hooked up to the suction filter, whereas underneath the two H.P. cylinders is the common 2nd stage discharge volume bottle hooked up to the after-cooler located horizontally at the grade level. Both the volume bottles were initially supported from the ground on separate pedestals and connected to the respective cylinders through adaptors.

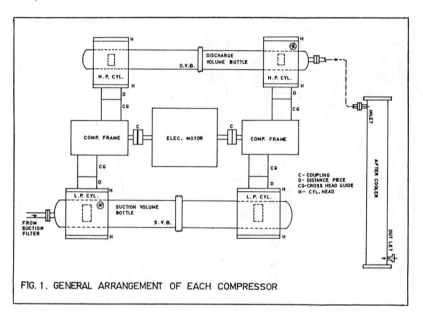

FIG. 1. GENERAL ARRANGEMENT OF EACH COMPRESSOR

The foundation block used for compressors is shown in Fig.2.

Misalignment of the shafts at a coupling is not permitted to exceed 0.1 mm $(4 \times 10^{-3}$ in) for trouble free operation of the machines. Also, it has to be ensured that the weight of the volume bottles located underneath the cylinders is taken by the pedestal supports and not by the cylinders, to which they are connected.

Due to large foundation settlements upto 60 mm (2.36 in) the blocks showed excessive tilts and frequent misalignments at the couplings, so that the machines could not be run for more than two years of their installation. Continued settlements and misalignments were apprehended, particularly under vibrations during running of the machines. The stability and even the strength of the soil was doubted. This paper reports the findings of a study undertaken to investigate the possibility of settlements in the future, and to measure as well as compute the vibration amplitudes of the foundation blocks based upon dynamic properties of the soil determined from in-situ tests on the foundation blocks (Prakash and Kumar,1984).

After the dynamic testing on one of the three compressor foundations its performance was monitored at different loads (Prakash and Kumar, 1984).

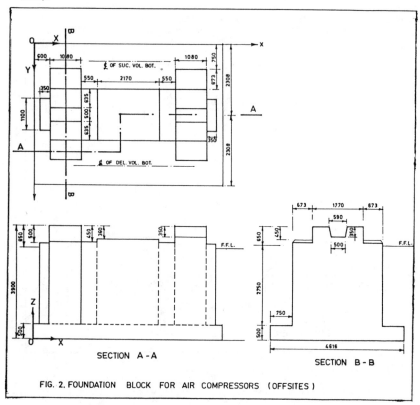

FIG. 2. FOUNDATION BLOCK FOR AIR COMPRESSORS (OFFSITES)

SOIL AND VIBRATION DATA

The soil data has been obtained mainly from the Dadina Report. The top 6 m (20 ft) depth of soil was excavated and replaced by fine sand conforming to Grading Zone III of I.S. Code 383-1970 (EIL Report, 1983). It is feared that this backfilled soil might have been disturbed due to inundation of the pit during the first monsoons when the filling had proceeded about half-way, causing degradation in the specified density of 90% of maximum dry density. The SPT N values were found to be less than 15 with an average of about 10. The water table was at about 1.3 m (4.3 ft) depth.

The second layer of soil, having 6 m (20 ft) thickness, consists of moderately firm bluish-grey clayey to sandy silt with an average N value of 11 and a maximum value of 30 at its base. It also contained about 3% mica. Typical ϕ and C values for this soil were 7° and 0.03 Mpa

(4.27 psi) respectively. The liquid and plastic limits were 38 and 19 and dry density 14.2 kN/m^3 (90.5 lb/ft^3). This layer is sandwitched between two water bearing sandy deposits.

The third and the lowest layer consists of dense to very dense grey sand to sandy gravel of maximum size 100 mm (4 in). N values showed a steep rise with increase in depth. The N value at 15 m (50 ft) depth from ground was more than 75.

Permissible single amplitude of vibration is 100 microns (I.S. Code 2974, Pt. I, 1982, Pt. IV, 1979, Major, Vol.III, 1980), which corresponds to permissible peak velocity of 6 mm/sec at the machine's operating speed of 595 rpm.

The unbalanced forces acting on the foundation block are as follows:

Moment in Y-Z (transverse) plane, M_x = 10.59 kNm(7.81 kip-ft)

Moment in X-Z (longitudinal) plane, M_y = 2.55 kNm(1.88 kip-ft)

Moment in X-Y (horizontal) plane, M_ψ = 12.08 kNm(8.91 kip-ft)

SETTLEMENT OBSERVATIONS BEFORE DYNAMIC INVESTIGATIONS

A total settlement of about 60 mm was observed between Feb 1981 and August 1983 in one of the three foundation blocks. However, only 4 mm (0.16 in) settlement occurred during the last one year of this period.

The centering of the cylinders was appreciably disturbed due to the settling down of the foundation blocks of all the three compressor units, which resulted in jamming of the adaptors between the cylinder nozzles and the volume bottles, and misalignment of the shaft couplings. Also the volume bottles under the cylinders had separate supports so that they could not sink with the foundation blocks, thus subjecting outer ends of the cylinders to upward forces (EIL Report, 1983).

TESTS ON FOUNDATION BLOCK B

Vibration tests were carried out on the middle compressor block between 24.11.1983 and 26.11.1983. The machine was completely dismantled from this block in order to carry out vibration tests and preloading tests. Settlement measurements with water-tube meters (accuracy ± 0.1 mm) were also recorded and continued during the pre-loading test upto 5.1.1983. A cross-check on the settlements was made by observations with a dumpy level which had an accuracy of ± 0.5 mm (0.02 in).

Two types of vibration tests were performed to obtain the natural frequencies of vibration of the block. These are free vibration tests and steady-state forced vibration tests. The dynamic soil constants could be computed from the natural frequencies obtained from these tests.

Free Vibration Test - Longitudinal (x-) Direction of Block

Free vibration tests wsere performed on foundation block B by applying a pull of about 60 kN (13.5 kips) with a chain pulley block and releasing this load suddenly by means of a special clutch. The load was measured with a dynamometer, which was replaced by a gap piece before releasing the load suddenly, in order to protect this instrument.

No free vibrations occurred under the 60 kN (13.5 kips) horizontal pull on account of the obstruction from the 150 mm (6 in) thick reinforced concrete floor slab abutting against the foundation block.

Free Vibration Along Vertical Direction

In the vertical direction also a pull, P, of 50 kN (11.2 kips) was applied and released suddenly to produce free vibrations. Reaction was taken from the 100 kN (22.5 kips) capacity overhead travelling crane in the compressor house. From the free vibration record the following information is obtained:

Natural frequency of vibration of foundation
block in vertical direction, f_{nz} =   10.43 cps   =   626 cpm

Damping expressed as a fraction of critical   =   5.7%
damping value, ζ

Other Data:

Weight of machine   =   206.5 kN (46.43 kips)

Weight of foundation block, mg   =   1321.6 kN (297.11 kips)

Base area   =   4.616 x 6.63   =   30.60 m^2 (329.4 ft^2)

The coefficient of elastic uniform compression of the soil, C_u, is thus obtained as follows:

$$C_u = 4\pi^2 \, m \, f_{nz}^2 / A$$

$$= 4 \times \pi^2 \times 1321.6 \times (10.43)^2/(30.60 \times 9.806)$$

$$= 18920 \text{ kN/m}^3 \ (120.5 \text{ kips/ft}^3)$$

Vertical soil stiffness, k_z = $C_u A$ = 579,000 kN/m (39,676 kips/ft)

Displacement amplitude, Δ_{st} = P/k_z = 86 micron (3.4 x 10^{-3} in)

Strain level   = Δ_{st} /B   =   0.86 x 10^{-4}/4.616 = 19 x 10^{-6}

The amplitude levels in free vibration tests are thus seen to be of the same order as the maximum permissible vibration amplitudes, so that the C_u value computed from free vibration test need not be corrected for strain level. Hence C_u equal to 18.92 MN/m^3 (120.5 kips/ft^3) has been adopted for theoretical analysis of vibration amplitudes under unbalanced forces.

Forced Vibration Test - Vertical (-z) Direction of Block

Steady State Forced Vibration Tests were carried out on block B with a Lazan type eccentric mass mechanical oscillator capable of providing maximum dynamic force of 1.52 kN (342 lbs) at 10 cps speed. The equivalent static force can be much larger on account of the dynamic magnification factor which attains a maximum value of $0.5/\zeta$ (= 8.77 for ζ = 0.057 in vertical direction) at resonance. The oscillator force increases quadratically with speed, attaining a value of 6.08 kN (1367 lbs) at 20 cps.

Two 12 mm thick mild steel plates with ribs were specially made to mount the motor-oscillator assembly on the foundation either at the central or end positions of the block. The overall sizes of these plates are 1500 x 1600 mm (4.92 x 5.25 ft) and 830 x 1690 mm (2.72 x 5.54 ft).

Vibrations were measured with a Miller-type accelerometer using a universal amplifier of high sensitivity and a direct ink writing oscillograph capable of recording accurately vibrations upto 70 cps.

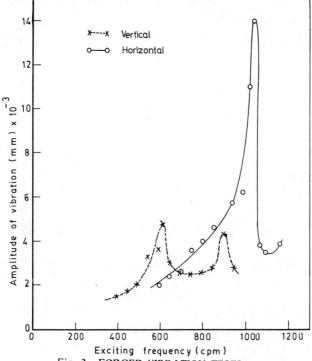

Fig. 3. FORCED VIBRATION TESTS

The natural frequency of vibration obtained in the vertical direction, f_{nz} = 620 cpm = 10.33 cps (Fig.3). This value is quite close to the earlier value of 10.43 cps obtained from the free vibration tests. Computations have been made as per I.S. Code 5249-1977.

Forced Vibration Test - Horizontal (-y) Direction of Block

With the oscillator mounted near the center of block, the foundation could be excited in the rocking-cum-shear mode of vibration. The second natural frequency obtained from this test is 17.33 cps (Fig.3) against the computed value of 14.95 cps (897 cpm). The first natural frequency is computed as 324 cpm. Both the natural frequencies are far away from the machine's operating speed of 595 rpm, hence the vibration amplitudes will be very small.

Settlements of the Block During Vibration Tests

Settlements were recorded after each vibration test. Practically no settlements were observed in the NW and SE corners of the block during the three days of vibration tests from 24.11.1983 to 26.11.1983. However, the NE end of the block went up by nearly 1.0 mm whereas the SW end settled by about an equal amount indicating a differential settlement of 2.0 mm (0.08 in) across the 6.260 m (20.5 ft) long diagonal, that is, a rotation of about 1 in 3000 in the diagonal direction.

Spirit levels indicated a maximum tilt of 1 in 4000 in the transverse direction and 1 in 4400 in the longitudinal direction.

No total settlement of the block was, however, observed.

Settlement of Block B During Preloading Test

Preloading of one of the foundations was suggested to ascertain the possibility of further settlements in future, the nature of settlements (whether instantaneous or slow consolidation type), and to qualitatively check the bearing capacity of soil from the rate of settlement with load. A total preload of 900 kN (202 kips) was applied which gives a net overload of 693 kN (156 kips) or 44% over the total working load on the block including the machine weight. The preloading was started on 30.11.1983 and completed on 22.12.1983. Thereafter unloading started on 27.12.1983 and finished on 2.1.1984.

The settlements during preloading proceeded as given in Table 1.

It is seen that the first 300 kN (67 kips) load causes no settlement. This is apparently due to the removal of 207 kN (47 kips) weight of the machine before preloading. Since the reading accuracy of the water tube meters is ± 0.1 mm, a maximum error in the average and differential settlements can be ± 0.2 mm. Hence the maximum possible averable settlement is 0.4 mm at 900 kN (202 kips) load and the accompanying maximum differential settlement could be 1.8 mm. The differential settlement somewhat increases during unloading. This may be due to eccentric placing of the pre-loads. The residual settlements after complete removal

Table 1 - Settlements in Preloading Test

Preload (kN)	Total Settlement (mm) SW End	NE End	Average Settlement(mm)	Differential Settlement (mm)
300	0.0	0.0	0.0	0.0
600	1.2	-0.4	0.4	1.6
900	1.0	-0.6	0.2	1.6
500	1.2	-0.6	0.3	1.8
200	1.6	-0.5	0.5	2.2
0	1.2	1.2	1.2	0.0

of the preload are not reliable due to disturbance in the water tube meter at the NE end of the block. The maximum settlement of about 0.4 mm at 900 kN (202 kips) preload, i.e., 690 kN overload (156 kips) overload is quite small and indicates that the rate of settlement with additional loads does not increase with overloading upto 44% of the design loads. Also the maximum tilt is 1.8 mm (0.071 in) in block width of 3116 mm (10.22 ft), i.e. 1 in 1700. This tilt is largely due to eccentricities in the preloading, so that under the actual machine loads much smaller tilt would occur. In practice, the tilt would be much less due to small machine weight of 207 kN (47 kips) compared to the load of 900 kN (202 kips) applied in preloading. In fact the machine is levelled initially so that whatever little tilt occurs due to machine's self-weight is automatically eliminated in the erection process. Moreover, since the total settlement in the soil is very small, there is little possibility of tilt during operation of the machine.

Further, it is seen that the total settlements take place more or less instantaneously, in about a day or so, under an applied load. This is normal behaviour of sandy soils.

Since the foundation block B showed no total settlements under vibration tests and very small total settlements in preloading, they were not likely to experience harmful total or differential settlements. Consequently, it was planned to carry out test runs on block A, monitoring the settlements and vibration amplitudes and checking the shaft alignments at the couplings.

The observed values of settlements and tilts during test runs were supposed to indicate if these did taper off in a reasonable period of time, say 4-5 days. The machine was to be first run at no load and only after a satisfactory 'no load' test, partial/full load test runs were to be conducted.

Repeated Vibration Tests on Block B After Preloading

The free and forced vibration tests on Block B were repeated to find out if any improvement in soil stiffness has occurred due to preloading.

The natural frequencies of vibration in this block were found to have increased by about 3.5 per cent so that the elastic coefficients and hence the dynamic stiffness of the soil showed an increase of about 7 per cent, the stiffness being proportional to the square of the natural frequency of vibration. Since this increase in the natural frequencies of vibration is marginal, the value of C_u already computed from earlier tests has been adopted in the subsequent analysis, being somewhat conservative.

No significant increase in natural frequencies or C_u after pre-loading also indicated that the soil has more or less stabilised against settlements. Had large settlements occurred during preloading, a significant increase in C_u could have been expected.

TESTS ON FOUNDATION BLOCK A

The machine was run on no load on 7.1.1984, 8.1.1984 and 9.1.1984 by first coupling the compressor on one side, then on the other side of the motor and finally on both the sides. Next run was taken on 11.1.1984 with the valves connected but without any load. Finally two test runs were taken on 12.1.1984 on half load and full load. Vibration levels and settlements were monitored throughout these tests.

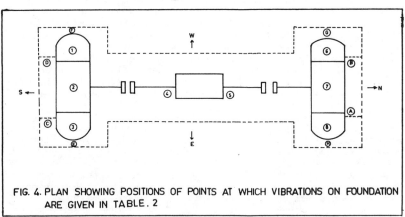

FIG. 4. PLAN SHOWING POSITIONS OF POINTS AT WHICH VIBRATIONS ON FOUNDATION ARE GIVEN IN TABLE. 2

The points at which vibrations were recorded with IRD Mechanalysis Vibration Meter is shown in Fig. 4. The measured vibration displacement and velocity amplitudes are given in Table 2. The settlement and tilt observations are plotted in Figs. 5 through 8.

Table 2 - Vibration Amplitudes on Foundation

| Load | Location of Observation Point (Fig.4) | | | | | | | | Direction |
| | Foundation Block | | | | Machine's Body | | | | |
	A	B	C	D	E	F	3	8	
1st Run No Load	2.0*/0.8	2.5/1.0	2.0/1.4	3.0/1.6	1.5/0.5	1.4/1.0	8.5/1.0	18.0/2.0	V
	5.0/0.9	4.0/1.0	5.0/1.4	8.0/1.6	8.0/0.5	3.0/0.4	6.0/2.0	5.0/1.4	H
2nd Run No Load	2.2/1.2	3.2/2.0	3.6/0.6	4.0/1.2	3.0/0.4	2.4/0.4	22.0/2.0	72.0/5.8	V
	4.0/1.5	8.0/1.0	7.0/1.5	7.0/0.8	6.0/0.5	5.0/0.4	50.0/2.5	50.0/3.4	H
3rd Run 5 kg/cm² Load	6.0/2.4	6.0/2.4	6.4/0.9	8.0/1.4	4.0/0.8	5.0/0.5	56.0/4.5	95.0/7.0	V
	8.0/1.6	9.0/1.4	7.5/1.6	12.0/1.2	6.0/0.7	5.0/0.6	120.0/5.0	120.0/5.0	H
4th Run 7 kg/cm² Load	6.0/1.5	5.5/2.6	6.0/1.7	10.0/11.7	4.0/0.8	4.0/0.6	40.0/4.4	70.0/7.0	V
	7.0/1.2	9.0/1.5	9.0/2.0	14.0/2.5	7.0/0.7	10.0/1.0	120.0/5.8	120.0/5.6	H

*Double amplitude in microns/Velocity in mm per sec

Observed Vibrations of Block A During Test Runs

From Table No.2 it can be seen that the vibration levels generally increase somewhat with the machine load. The vibration levels in the foundation block for half load are about 50% higher and for full load about 100% higher than at no load.

The maximum recorded velocity of vibrations at full load is 2.5 mm/sec (0.1 in/sec) which is much less than the permissible value of 6 mm/sec (0.24 in/sec). Also the maximum double displacement amplitude is 14 microns (0.55×10^{-3} in) in the foundation block and 120 microns (4.7×10^{-3} in) on the machine. These too are smaller than the permissible value of 200 microns (8×10^{-3} in). Moreover, no misalignment occurred during the entire period of five days on which test runs were conducted between 7.1.1984 and 12.1.1984.

Observed Settlements of Block A During Test Runs

The settlements were measured with water tube meters at all the four corners of the foundation block at a transverse spacing of 3116 mm (10.22 ft) and longitudinal spacing of 5430 mm (17.81 ft). In addition, four sensitive spirit levels were also installed on the block on all its four sides to directly measure the tilt and compare with the values computed from water tube meter observations.

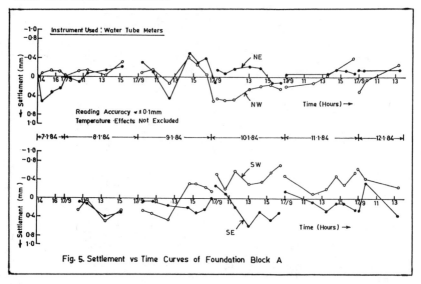

Fig. 5. Settlement vs Time Curves of Foundation Block A

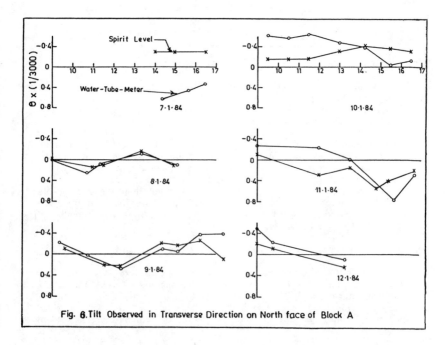

Fig. 6. Tilt Observed in Transverse Direction on North face of Block A

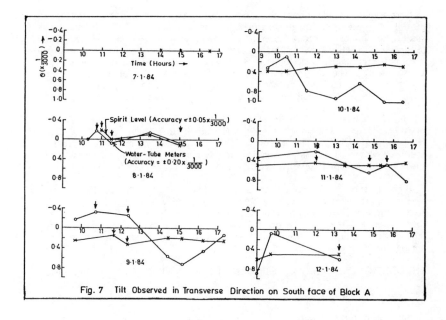

Fig. 7 Tilt Observed in Transverse Direction on South face of Block A

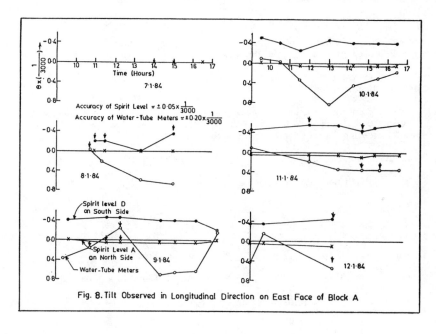

Fig. 8. Tilt Observed in Longitudinal Direction on East Face of Block A

Settlement observations obtained from water tube meters are given in Fig. 5. Significant temperature effects are apparent from these plots since the values fluctuate about the mean position without showing any progressive settlement. This is typical character of the temperature effects which caused movement of the fixed reference end of the water tube meter. The South-West end of block shows an uplift of 0.35 mm whereas the South-East end shows a settlement of 0.15 mm, thus producing a small tilt in the transverse direction at the south face of the block.

The tilt observations from spirit levels shown in Fig. 6 through 8 are more consistent and much less erratic since they have relatively little temperature errors. The tilt values from water tube meter observations are also plotted for comparison in these figures. These are seen to fluctuate about the spirit level curves.

Figs. 6 through 8 show that a negligible tilt of 1 in 30000 occurred in the longitudinal directon on the east face and a tilt of 1 in 6000 (relative settlement of 0.5 mm) occurred in the transverse direction on the south face of Block A during the no load tests whereas no tilt occurred during the partial and full load tests. Also no total settlement occurred during any of the test runs.

THEORETICAL NATURAL FREQUENCIES AND VIBRATION AMPLITUDES

Vertical Axis - z

Vertical natural frequency, f_{nz} = 582 cpm

Natural frequency in yawing, $f_{\psi z}$ = 578 cpm

Since the machine's operating speed is 595 rpm, there is possibility of building up of vibrations in the yawing mode, i.e. rotation about z-axis.

No resonance can occur in the vertical mode since there is no exciting force in this direction.

Damping in the rotational mode could not be measured. Thus a design value of $0.4/2\pi$ = 6.37% (Major, Vol. III, 1980) has been used which yields an amplification factor of 7.85 or approximately 8.0

Exciting moment M_ψ = 12.08 kNm (8.91 kip-ft)

Torsional stiffness about x-axis,

k_ψ = $C_\psi \ J_z$

J_z = $I_x + I_y$ = 54.341 + 112.105 = 166.446 m^4 (19285 ft^4)

C_ψ = 1.5 C_τ = 0.75 C_u = 14190 kN/m^3 (90.3 $kips/ft^3$)

k_ψ = 14190 x 166.446 = 2361,870 kNm/rad (1742,110 kips-ft/rad)

ψ_{st} = M_ψ /$k_{\bar{\psi}}$ 12.08/2361870 = 5.11 x 10^{-6} rad

$$\psi_{dy} = \psi_{st} . \mu = 5.11 \times 10^{-6} \times 8 = 40.9 \times 10^{-6} \text{ rad.}$$

Displacement amplitude $= \psi_{dy} \times \dfrac{L}{2}$

$$= 40.9 \times 10^{-6} \times \dfrac{5430}{2} = 0.11 \text{ mm } (4.3 \times 10^{-3} \text{ in})$$

However, the floor slab abutting against the foundation block imparted considerable stiffness to the block, thus averting resonance. Consequently, the measured vibration amplitudes due to yawing motion were much smaller than 110 microns (4.3×10^{-3} in), the maximum recorded value of single displacement amplitude being only 7 microns (0.3×10^{-3} in).

Transverse Axis -y

The unbalanced moment about this axis has a small value of 2.55 kNm (1.88 kip-ft) whereas the stiffness of the subgrade is largest in the longitudinal direction. Hence the vibration amplitudes obtained are insignificant being less than 1 micron (0.04×10^{-3} in).

Longitudinal Axis -x

The unbalanced moment about the x-axis would produce displacement amplitude of less than 10 microns (0.4×10^{-3} in).

Thus the theoretically computed vibration levels are within the permissible limits in all the three directions.

CONCLUSIONS

1. Though initially very large settlements upto 60 mm (2.36 in) were observed on the foundations, no total settlements were subsequently observed. The estimated value of settlements using charts of Terzaghi and Peck is 32 mm (Terzaghi and Peck 1967) and the elastic settlement 6 mm (Prakash, 1981). The preloading test further confirmed that the settlements had stabilised.

2. No total settlements were observed in block A during trial runs of the compressor as well as in block B during the free and forced vibration tests.

3. No misalignment was observed in the shaft of the machine across the couplings during several test runs of the compressor on block A, including test runs at half and full loads.

4. A maximum tilt of 1 in 6000 occurred in the transverse direction of block A during test runs at no load, yielding 0.5 mm (0.02 in) differential settlement across the block's total width of about 3.0 m (10 ft). This tilt did not increase during further test runs at no load with valves, half load and full load. No tilting occurred in the other directions.

5. The measured velocities as well as displacement amplitudes on block A are quite small and well within the permissible limits, even at full load. The theoretically computed values are also safe.

6. The volume bottles should preferably be supported from the main foundation block, otherwise a flexible connection provided at the adapters. This will provide long term protection to the problem of misalignment. Occasional checking of the shaft alignment might be required in the first six months as a precautionary measure.

7. All the tests indicate that the compressor foundations are quite safe against vibration amplitudes and settlements. Thus the chances of tilting of foundations and misalignment of shafts in future are very remote. However, the performance of the foundations of the other two compressors may also be monitored during their test runs to ascertain their behaviour.

APPENDIX-I. - REFERENCES

1. Arya, S., O'Neill, M., and Pincus, G., "Design of Structures and Foundations for Vibrating Machines", Gulf Publishing Co., Book Dn., Houston, 47-54, 1979.

2. Dadina, K.N., "Report on Soil Investigation for Polyster Fibre, Xylene and DMT Plants, BRPL", 120-135, June, 1980.

3. Engineers India Ltd., "Report on Foundation Settlement/Misalignment of Air Compressors (Offsites) of Petro-chemicals, BRPL", Aug.1983.

4. I.S. 383, "Coarse and Fine Aggregates from Natural Sources for Concrete" 7-11, 1970.

5. I.S. 2974 (Pt.I) - "Foundations for Reciprocating Type Machines", 10-11, 1982.

6. I.S. 2974 (Pt. IV) - "Foundations for Rotary Type Machines of Low Frequency", 15-16, 1979.

7. I.S. 5249, "Determination of Dynamic Properties of Soils", 1977.

8. Major,A., "Dynamics in Civil Engineering, Analysis and Design", Vol.III, Akademici Kiado, Budapest, 94-101, 1980.

9. Prakash,S., "Soil Dynamics", McGraw Hill Book Co., 110-111, 1981.

10. Prakash, S., and Kumar, K., "Final Report on Dynamic Analysis and Settlement Investigation of Foundations for Air Compressors (Offsites) at BRPL, Bongaigaon", Consultancy Project Report of Central Building Research Institute, Roorkee, India, No.SE/CONS-72/83-84, April, 1984

11. Terzaghi, K., and Peck, R.B., "Soil Mechanics in Engineering Practice", 2nd Edn., John Wiley, 490-491, 1967.

SUBJECT INDEX
Page number refers to first page of paper.

AUTHOR INDEX
Page number refers to first page of paper.